The Autonomous Economy

Book I

The Strong Man

OAKSHADE™

Dedication

This book is dedicated to my family, my wife Angela and my son Michael, both of whom helped inspire it.

At various occasions throughout the years, both of them served as sounding boards for me to bounce thoughts off, allowing ideas to gel in my head. Without their support and patience while forming the ideas which inspired the book, along with their thoughtful suggestions, it would have turned out considerably less palatable.

CP McCollum

Acknowledgements

I would like to recognize the following folks with a Special acknowledgments for their valuable feedback and helpful suggestions:

My wife Angela, my mom Judy, Joan Dieschbourg, David Sanford, My son Michael

Jay Josephs, owner of Rolli Porkloin of Melbourne Florida

A very special thanks to friend, futurist visionary, conceptual artist, and author -- Syd Mead of Syd Mead Inc., for his invaluable and insightful feedback.

Forward

** Warning Spoilers **

I started developing the outline for this book in 2006. At that time there was very little on the public landscape discussing an autonomous future, with the exception of some silly science fiction movies, depicting intelligent robotic machines taking control of the world with the aim of destroying humankind, which represents a threat to their survival.

Some may speculate on the question of why it took 9 years to write and publish a book. The answer involves many factors including that my original intent was to publish only the one book, which grew in scope and complexity, so I decided to split it into 3, 2 of which are currently at only partial stage of completion. Then, I decided to take a year and write 2 others, both of which are non-fiction. The first of those I published about a year ago, after printing a limited preview (Advance Reader Copy) supply of this book in the summer of 2013. I decided to take a break and work on the other books, due to the timeliness of the material, believing I had time later to complete and publish the 1st in *The Autonomous Economy* series.

At the time I started toying with the idea of writing anything at all, there was very little on the horizon or in current print in a similar vein; the humanoid robots taking over our lives theme; so it seemed I had plenty of time. This is my first work, so forecasting and measuring the work is daunting when you have little experience. I discovered everything the hard way. Not being a prolific reader myself, I discovered that editing and structure is murder; and that research is tedious and expensive. Although I have never been lacking in material and ideas, having never taken a course on creative writing, nor ever having read a book on the subject, I had to experiment with and learn how to convert long descriptions of scenarios or events into scenes with dialog, then structure all of the many disjointed stories for as seamless a flow as possible. When it comes to character development...? Forget it..., not high on my list of mastered ability..., yet. I had to do my own graphics and editing, researching and learning authoring tools, publishing resources, etc. It was a learning experience to say the least.

Since I began the project, much has changed in the popular media regarding the advent of an autonomous world. After having written a great deal of the manuscript I began to notice more mention of the material content of this book start to emerge in news reports and popular media, to the extent that I realized some of it was particularly close in scope and tenor to the content of my uncompleted work. I began to see direct depictions in movies and news accounts of scenes that on the surface, seemed to be right out of my manuscript. I thought about it and decided the phenomena must stem from, and occur according to the idea of *'Art imitating life'*, or is it *'Life imitating art'*, I'm confused as to which one imitates which; or maybe its that each inspires and influences the other.

I think that the phenomena is subjective and to be sure it certainly involves perception which may be cognitive distortions, or it may stem from several other factors including outright plagiarism. The perceptive cause may indeed apply to my case. As such it might apply when one does not notice things which normally occur around them until they become involved with whatever the subject is. As an example, one may not notice how many of a certain model of automobile are on the road until after they have purchased one, then they notice that the same model is everywhere. The same models have probably been there in numbers, just not noticed by them until they become involved with it. That explanation is hard to believe in my case though because I was involved with the subject matter for several years before I began to notice any significant advent of it in the popular media.

That explanation should have satisfied my perplexed mind except for the fact that I outlined several chapters where afterward I discovered recent, unusually close published content, which appeared after having written mine. As an example, I outlined the chapter titled, *'Bot V Bull',* in late 2010, and wrote most of the body in early 2011. In late (October 2) 2011 a film titled 'Real Steel' was released in the US which my wife and I went to see and were very surprised to see that within the first few minutes there were numerous scenes depicted with close similarities to the material in my book. In the book, the 1st 'robot' depicted is a hulking industrial robot in Japan; in the movie the 1st 'robot' is a hulking 'boxer' robot in a box from Japan, which could be coincidental, but the next scene of the movie has a robot wrestling a bull at a rodeo corral setting in Texas, while in this book, the 'robot' is debuted at a rodeo in Texas, where it rides a bull. Both scenarios are uncomfortably close with both settings in Texas. I was taken aback and began to worry about being labeled a plagiarist if I did not alter the material. I decided to leave it the way it was because I have not plagiarized anything and I can document when my material was created as proof, which was several months prior to any knowledge of the film by me, at a minimum. If anything, in my mind, if plagiarism has indeed occurred, it was in the reverse direction. I will vigorously defend the originality of the material in this work.

It is difficult to imagine the glaring similarities to be the result of some sort of coincident, but still I had no explanation for it to be anything else. That is when I decided it must be *'Art imitating life.'* However, I also began to wonder if it was possible that someone may have penetrated my local network, hacked my computer and stole my manuscript, but it is very difficult to believe that to be possible or to imagine what motives could be involved. How could such a scenario occur? I am generally technically savvy, and take precautions against security risk. I wondered, who could possibly be interested in my work; I'm an unpublished author working on my first book; who could even know there is anything worth reading?

Still, I believed it must be some sort of coincidence because of the obvious problems with the alternative explanation, however, that was in the face of an ever mounting number of incidents of the same type which I noted and compiled over a few years. I was not satisfied, so a few years later I went back and reviewed the movie again, and was shocked to find an even more, inescapably close similarity I had previously overlooked. In the movie, the robot fighting the bull did a mid-air acrobatic flip, pushing itself off the back of the bull with one foot during the contest, which was the exact scene my book described. In my book, the robot dismounts with a flip, pushing itself off the back of the bull with one foot. This is way too close to simply be a coincident. Now I am not sure what to think.

It is not just these few incidences, but there were many, many more examples of *'Art imitating life.'* In the later part of the book, I described a scene of mass protest where people in many cities all over the world occupy large tracts of public property in protest. Within a few months after describing this scene, the *'Occupy Wall-street'* protests erupted in the news, where many people in many cities did the same. Again, is this art imitating life? If it is, it is in the reverse order.

I had outlined the advent of autonomously driven automobiles as far back as 2007, and incorporated a great deal of the material in the manuscript. Since I researched the sparse material available and become minimally aware of the ideas involved, going back into the mid 2000's, I have had many discussions about the subject with my then young teenage son, who is now 24. Again, at the time, there was nothing on the Radar about the subject, and no one I knew had heard anything about it. If I attempted to discuss the idea with anyone, I would get comments like, *"Oh, you mean like the Jetsons?"* Within months after the detailed writing of the chapters which I had outlined many years prior, news reports started 'informing' the public that there was an autonomous automobile in their future, something which I imagined to be at least a decade away from any public disclosure.

Then there is the mass Mexican invasion of the US over the Texas border, with the white-house deciding to stand down and allow it, described in the section titled *'The Sub Standard Labor Solution'* in the second chapter, and continued in the chapter titled, *'Simple State of Siege.'* It arrived in the flesh only a few weeks ago, in July 2014, just prior to me deciding it was time to write an update to this forward. The list of 'coincidences' I have given here is only a sample of the actual number I have been made aware of over the last 4 to 5 years, which are probably the most glaring in scope.

In the chapter titled *'Artificial Stupidity'*, I outline a scene where a professor is teaching a class and discussing the subject of artificial intelligence and robot training in the context of deterministic vs non-deterministic operation. He picks up a rubber ball, bounces it, then tosses it to a student who catches it, to make a point. In the movie titled, *Knowing*, with Nicolas Cage, an extremely similar scene unfolds in the same setting of a classroom with a professor discussing determinism, who picks up a globe and tosses it to a student who catches it. In the case of the book, the events and context are material to the story, while in the movie they seem to be completely unrelated to the story.

Also the chapter *'Artificial Stupidity'* describes the phenomena of how machines must be taught virtually everything about the world from scratch just as humans are. In my book, the machines are called ARCHIEs', and CHARLIEs', while in the movie Chappie, the same baby-steps theme is present, but the machine is called *'Chappie.'*

"He's a happy chappie; lets call him Chappie"

As in the movie, Chappie, there are other parallel themes involving the use of humanoid robots for law enforcement, having been outlined in early 2011, which are forthcoming in subsequent publications in *The Autonomous Economy* series.

The clincher was a few months after writing the two chapters titled, *'Auto Fleet'* and *'Cloud Self.'* In these, I describe the total loss of all personal privacy due to a criminal global government regime instituting massive spying on the populace, to the extent that they can control all media, then eavesdrop on and get detailed information on every aspect of every person on the planet, making it virtually unavoidable; something which I knew was theoretically possible, and which I suspected was occurring, or would occur to some very limited degree. My knowledge of such a subject goes back to my work in the late 1990s involving data-mining and *'key profile-compilation'*, which culminated in a foundational USPTO patent no. 6,594,691 titled, *'Method and system for adding function to a web page'*, and its many derivatives; the purpose of which is to facilitate obtaining and tracking information about an anonymous internet user. The user is not personally identified, but their profile is something which can identify their internet presence through the breadcrumb trail of behavioral data left behind.

I was again shocked when within a year and a few months after writing the two chapters mentioned, news broke of whistleblower Edward Snowden, who spilled the beans that the NSA (National Security Agency) of the US government had pretty much built and was illegally operating such a network and mechanism for many years prior. Again the question, is this *'Art imitating life'* or maybe the other way around; or is it just a huge coincident, or something more? Even if my manuscript had been stolen, it could have absolutely no bearing on the establishment of the US government's global spy network, which took decades to put in place, except that, to consider the remotest stretch of credulity; if certain elements within the government obtained my manuscript, and saw similar speculation emerge from other contemporary sources, could that constitute a signal to those keeping the secret that it cannot be kept much longer, in-which case an official 'controlled leak' is triggered in order to get ahead of it in attempt to control dialog and perception?

I am not suggesting that any of these things are what occurred; it is pure conjecture. I am merely pointing out the extreme degree of correlation between events, the probability of which I personally put considerably beyond coincidence. On the surface, the phenomena can only seem to reflect that some of my ideas and materials are influenced by news of current affairs and events, some of which may precede larger public dissemination, without my being aware of it's exact origin. The details of what I have written are my own ideas, although I have to admit that I am influenced by the current paradigm and events, and maybe considerably more than I am actually aware. While I never undertook a concerted research effort prior to writing the material, but only drew upon my knowledge and experience of what was possible, and with considerable imagination, the spying scenario which I described in the chapters is strikingly similar in scope to what actually exists, and it was written well before any public knowledge of its existence surfaced.

Snowden's revelation led to the discovery of NSA spyware which predated public dissemination of the viral computer *worm* called Stuxnet, which infected and disrupted industrial controllers around the world, and is believed to be responsible for destroying some of Iran's uranium enrichment centrifuges. It was later determined that Stuxnet may have been a variant of Regin, and a self replicating and self spreading worm called Fanny, both of which were developed by *'The Equation Group'*, inside the NSA. These worms are the genesis of a family of such malicious code which have been widely embedded directly into the firmware of commonly sold hard disks and other hardware, throughout the world, going back as far as 2001.

With Snowden's revelations and subsequent discoveries, I could now find a plausible reason to support a scenario where my manuscript had actually been hi-jacked from my computer. His revelations detail how the NSA routinely hacks into private

The Autonomous Economy

networks, through established 'back doors' that are known of and facilitated by the manufacturers of telecom and networking equipment. It has since been revealed that the NSA records virtually all telephone conversations, and 'sniffs' out virtually all e-mail and text traffic conducted over the internet. While that is not proof of the hi-jacking of my manuscript for the purpose of selling it to the highest bidder, something which is like what has been alleged as commercial espionage by several that have looked into the activities of the NSA spy network, it does make the possibility much more plausible now than before.

This work is a historical account in every sense of the meaning. It deals with events which may never happen, so, for the most part it is fantasy. No one can accurately predict what is going to happen next week or next year, let alone 50 to 100 years hence, so it is fiction, albeit with some amazing similarities. In its creation, there was a conscious attempt to minimize the tendency toward sensational absurdity, and instead describe more a sober assessment; much of what are plausible events and circumstances driven by the dictates of human nature. Note that when history is described in advance, it has a path to follow. This work recites a historical account spanning a little over a century and a half, however it was written in advance!

C.P. McCollum, June 2015

About the Author

Mr McCollum is an electrical engineer by background and has been an independent product developer and inventor. He has a number of inventions and patents to his name involving computer software technology. He has been a business founder and entrepreneur in the technology field with areas of practice in engineering and electronic design. His background has given him the insight to depict much of the imaginative technical descriptions and themes in the story. The rest is mostly a result of vivid imagination.

Additional areas of interest and personal study cover ancient history and the origins and rise of the peoples of the earth, economics, and political systems. Although having no formal training in economics, the basic principles are simple and simply understood by anyone willing to learn.

He is currently an author, having published 2 books, with an additional 3 in progress. One of his books is titled 'The Sovereign Democratic Majority: Reconquering the American Frontier', which defines his political views as Sovereign Democratic Majority. He is married for 34 years with one adult son, currently living in Florida. He is an observant non-affiliated Christian protestant, having read and studied the Christian bible.

About the Book

The book came about as part of an idea for product development involving trends in computer technology in about 2001. It must be noted that some of what inspired the ideas put forth were discussions I had with my then 10 year old, very bright and inquisitive son, Michael. Several other discussions ensued between him and myself for several more years. I remember standing in line for a ride at Disneyland in Florida with Michael, then 14, and his mother, engrossed in deep discussion about how things might transpire, as the premise of the book emerged. I started the outline in 2006 as time permitted, and after considerable research into authoring tools, I began writing in earnest in late 2010.

BTW, I use Scrivener for handling the considerable manuscript, which it does beautifully. The book quickly grew in size to about 150,000 words, when I decided to split the story into a trilogy of books due to the time it would take to complete, and the shear size, expected to approach 400,000 words.

Although I do enjoy science fiction in movies and literature, I am not what some would consider 'Hard Core.' The inspiration is less from the essentials of the cyber-future genre, and more from politics, philosophy, and technology trends. The story in my estimation, is less about robots and technology, and more about people, and the ultimate destiny of humanity. It is my belief that the future, which will invariably involve the ubiquitous proliferation of humanoid machines, is less driven by technology, and more by human behavior. Robots and technology in general are byproducts of that which drives human beings, so the book is about us humans; people; what we are, where we have been, and where we are going.

Disclaimers!

* All characters appearing in this work are fictitious. Any resemblance to real persons living or dead, is mostly coincidental.

* The book is intended as fiction and for the purpose of entertainment, inspiration, and education. It makes no attempt to 'predict' the future so should not be taken as such. The book, and series is written in a flashback narrative format (convolute), as a chronological account of future events, describing and reciting the historical account from the perspective of the narrator, while foretelling it from the perspective of the reader. It was conceived and written with a conscious effort for pragmatic plausibility, and to avoid the impulse toward sensationalism and absurd fantasy. The process comes about by using thesis and inductive estimations of plausible future events, based on historical precedent; the motivating factors of human nature; and the 'follow the money' theory, which drives individuals, groups, and civilizations.

* There is no attempt to establish or interpret theological positions or doctrine. Related content is simply a fictional interpretation, and a mechanism borrowed to tell the story. I have no training in theology, but only an interest. I make mention here to deter those of the Christian persuasion who may like to accuse me of changing scripture. There is no change, and none was intended, but the original is preserved. The story is expanded by use of commentary and narrative. As mentioned elsewhere, I am a conscientious Christian believer.

C.P. McCollum

Other books by CP McCollum

Fiction

- The Autonomous Economy
(Three book series available in print and ebook at Amazon and elsewhere)
- Book I - The Strong Man *
- Book II - Edenopolis - Available - estimate 2 – 3^{Q16}
- Book III - Abdication - Available - estimate 1 – 2^{Q17}

Non Fiction

- The Sovereign Democratic Majority: Reconquering the American Frontier *
- Retire US Debt Through Substitution - Available - estimate 2^{Q16}

* Currently available in print and ebook
All books in print and ebook versions are available at amazon.com and elsewhere.

Contact or to Purchase:

Tel/Fax: (206) 202-3855
Email: info@sovereigntyproject.org
Web: http://www.sovereigntyproject.org
Web: http://www.oakshadeLLC.com

Notable Quotes

The kings of Europe would dare challenge us? We throw them the head of a king!
- Georges Jacques Danton
President, Committee of Public Safety during the French Revolution, 1793, contemporary of Maximilian Robespierre.

Give me control of a nation's money and I care not who makes it's laws.
- Mayer Amschel Bauer Rothschild

Since I entered politics, I have chiefly had men's views confided to me privately. Some of the biggest men in the United States, in the field of commerce and manufacture, are afraid of somebody, are afraid of something. They know that there is a power somewhere so organized, so subtle, so watchful, so interlocked, so complete, so pervasive, that they had better not speak above their breath when they speak in condemnation of it.
- Woodrow Wilson

The supranational sovereignty of an intellectual elite and world bankers is surely preferable to the national auto-determination practiced in past centuries.
- David Rockefeller

"If the American people ever allow private banks to control the issue of their money, first by inflation and then by deflation, the banks and corporations that will grow up around them, will deprive the people of their property until their children will wake up homeless on the continent their fathers conquered."
- Thomas Jefferson

Those who create and issue money and credit direct the policies of government and hold in the hollow of their hands the destiny of the people.
- Reginald McKenna, former Chancellor of Exchequer, England

CHAPTER ONE

Toward Singularity

The Peoples Chipper

Two men are looking out from a sandy scrub desert hill about a quarter mile away from an exciting event as a coyote might view it from that vantage. They had arrived at a transport depot moments before, then walked up the hill to observe the spectacle. While walking up, they catch the first glimpse of what is unfolding. A large web of tubes like what would be found at water parks, where one gets into the tube and slides down. Round and round, curving, spiraling down as it drops a considerable distance. The tubes are perched high on a bluff and weave and turn for about a quarter mile, dropping about 200 feet vertically. There appears to be about a half dozen of these 'rides' with long lines of participants. The scene is reminiscent of an amusement park with the joviality of a summer afternoon's fun. The two spectators observe the action at the 1st of the 'ride' events. Three men are tending the ride participants, only in this case, the participants are bound and the three men are helping them into the ride clearly against their will.

The first man, clutching and speaking into a microphone suspended from above like an arena announcer, welcomes the next bound man in line who appears to be around 70 years of age, "Welcome senator Schuman, we are honored by your participation", two men behind laughing and taunting. He then mockingly asks, "Tell us senator, what is your position on the current crisis?"

The man offers the senator the microphone, and he pleads, "This is insane, you can't do this, what kind of animals are you, please?", the

1

senator, weakened and trembling by debilitating fear, is overcome and drops to his knees begging, "Please don't do this?"

The man holding the microphone, follows the senator's mouth down as he collapses and asks, "What's that Senator, I'm afraid we didn't hear you?" The Senator replies in a gravely shaking, and whimpering voice, "please don't do this, I'm begging you?"

The man mockingly repeats asking, "Please senator, we are having trouble hearing you, you sound feeble, speak up man, I'm afraid we still can't hear you, please speak up?"

The Senator then tries to stand back up but is too weak, but he musters a much louder and more clear but still trembling voice, and loudly repeats "I said…, please…, don't…, do…, this!"

The man tauntingly replies, "Come now senator Schuman, this will be the ride of your life, what few seconds are left", snickering and mumbling under his breath as he turns his back on the prostrate and trembling man that has taken on a very ashen faced appearance. Death shows before the body knows.

A rousing loud cheer rises up as the three men grab the bound and whimpering senator and feed the now screaming man head first into the entry of the tube and push him on his way. Looking up, our two hill-top observers see a large lit video monitor board above, which displays the action and spells out the senator's name, 'The honorable Senator Schuman'.

Meanwhile, the first man reaches up and grabs the dangling microphone to offer commentary, "There he goes, that's probably the first time the senator spoke the truth, or said anything of substance. Finally Senator Schuman, we believe what you are telling us!"

A clearly audible ditty of circus music followed by the funeral dirge begins to play and the crowd cheers.

Our observers are watching intently so they follow the tube along its path while the senator, audibly screaming, flies at high speed down the tube and then draws to close on its destination which appears to be a small building like a garden shed with a sizable crowd of people standing in front of it in ecstatic and heightened anticipation. The shed has a sign on the side spelling out in large illuminated letters, 'Peoples Chipper.'

As the good senator arrives at the shed at high speed, there is a quick and ground shaking, pounding, blood curdling thud of impact, with grinding and breaking, like that of a commercial saw mill, combined with a two train head-on collision, happening simultaneously, unmistakably informing even the casual observer of the gravity that has befallen the participant.   A sizable amount of blood and bone tissue issues forth at high speed, a considerable distance toward the audience, many of whom are completely naked, and have the appearance of those having risen from the abode of hell, displaying extreme body modification, and the crazed look of the insane.   They run up to the disgusting issue, rolling in it, many of them covering themselves in its loathsome essence as they laugh hysterically and applaud, and smear themselves, and wallow around in the nauseating goo, like the animals the honorable senator called them seconds before.

All eyes return back to the top as the next man is brought forth up the steps to the platform, the next inline for the ultimate ride. This man is portly, very rotund, and protests his incapacitation and bindings as he struggles puffing and grunting while he is escorted up the steps.   The three men then grab the 400 plus-pound man, and grunt and strain as they push him into the tube. Just as before, the same litany of narration, commentary, diatribe, and angry mocking banter between the executioner with the microphone, and the victim ensue.   They give him a sending shove and down he goes. As he rounds the 4th or 5th curve, he suddenly lodges in the tube and will not move.   Focusing in on the horrifying scene, the fat man laughs hysterically and very loudly, relieved at his apparent reprieve.   His laughs then turn to shrieks of terror, and hysterical wails of weeping once again, as he realizes that it is only short lived and the realization of inevitability sets in again, so he screams and cries the guttural mournful howl of one deeply sorrowful and repentant for their misdeeds, interspersed with the Lords Prayer.

The crowd is audibly and visibly very disappointed and irritated, which they show by their own deep mocking howls of mournful crying and chanting. They look to each other running back and forth with the grimaced body expression of question, which they gesture to each other as if to express umbrage at being robbed of their entertainment. Observing back atop to the three men, they appear to be discussing what to do to remedy the problem.

The Autonomous Economy I - The Strong Man

The third man barks, "I told you, no more fatties, they clog every time, now the crowd is upset…, just move him along…, and send the other lard asses to the log roller!"

The first man then issues an order and signals with a gesture as if to say, "Get in there and unblock it!"

Then a small machine resembling a metal crab scampers quickly up and into the tube and down it goes. The tube is illuminated from within, so the crab can be seen in silhouette as it descends to the fat man. Observation by one having the vantage of being inside the tube, is witness, whereupon the machine is busy moving very quickly around the site analyzing the situation with obvious keen intelligence. As the man's nervous laugh turn to shrieks of terror, the crab unleashes a large array of powered cutting tools and proceeds to process the various limbs and mounds of flesh that seem to have induced the problem. It then fires a rocket to propel the screaming bloody quivering mass, and itself the rest of the distance. As it hits the chipper, the familiar stream of gore pours forth onto the jubilant crowd followed quickly with a shower of sparks and metal shavings.

Again, observing from a distance our two witnesses are walking the crest of the hill, and as they stroll, they face the arcade to their right. They then stop and turn and one of them points to the right. He watches from an angle that reveals what could be seen clearly from the vantage, which appears to be 5 or so more blood sport arenas, their names, and the spectacle of a huge crowd of tens of thousands that stretches well over a mile; Body-Peeler', 'Mana-Pult', 'Head Popper', 'The Dragon' and so forth. It does not end with these, but it's clear that this spectacle goes on across the desert, farther than one can see.

Our two observers remain standing and talking, still watching from a distance of about 50 yards. They turn and retreat back the direction from whence they proceeded. As they walk, behind them is the spectacle of a human head flying through the air about 200 yards to the crowd below, many of whom hold large baskets used in attempt to catch the grizzly object on its trajectory. As they walk, all around them is the bustling noise of a crowd following after them, which is not visible but surrounds them on all sides, seen only in silhouette and shadow as if they are in the midst of a crowd, yet they walk in total solitude.

Both men have very long dark full beards and are wearing long flowing but highly decorative robes like those of medieval monks, one white, the other grey. The second man, Peter, then turns to the first man, Michael Dalgleish, who is wearing a white robe, and asks, "I'm disgusted Michael..., how has it come to this?", shaking his head in disbelief.

Michael then begins to speak and tell the story as he has come to know the answer to the question perplexing Peter. They re-enter the vehicle they arrived in while in obvious deep discussion. The sun is just setting as they drive off engrossed in discussion.

Ants

Laying belly-wise on his made bed, hands propping his head up. He stares and his mind wanders while watching and wondering at the busy lives of the little critters, because he is a curious and smart boy. He thinks to himself; do they have brains large enough to know what they are doing, or is it just instinct that shows them? For hours he stares and marvels at the ant house that has been his prize since his 10th birthday just 6 weeks earlier. The ants have built an incredible structure since he put them into their sand filled enclosure. He added a few twigs of straw and some leaves to their house to see what they would do with it. They probably had not met each other before, but now they have started to work together and create something magnificent. It now resembles a small ant civilization complete with sloped tunnels going down, connecting a network of rooms and mounds that reach a considerable height from the original flat featureless surface atop the sand. The features are visible from the outside because the whole civilization is contained between 2 sheets of plastic glass about an inch and a half apart. The leaves have been cut into smaller pieces and along with the twigs, they have been moved to a corner and somehow glued together in a kind of makeshift bridge to the sky. The ants are trying to escape their city.

His concentration is interrupted by the loud voice of his mother Celia, "Michael, its time for dinner..., come on down now and set the table for me please."

Michael replies, "OK mom, I'm coming." Michael continues to linger where he is. He wonders, "Will they all die someday; will they be mad at me; will they get out of the glass in the night, and bite me while I'm asleep?" He has been thinking about what would happen if he gave the colony a good shake.

Michael Dalgleish

The vehicle Michael and Peter are in arrives at a large civic hall in a metro downtown area, where they both enter. The crowd is very large, estimated to be 10,000 or more, excited and waiting with a raucous tone. They are already gathered and appear to have been there for a considerable time, coming and going. The event is an open forum which has been ongoing continuously for a long time. Peter makes his way into the crowd, while Michael goes up to the stage area where he continues the story. The crowd, anticipating his words, quiets down as he begins to speak.

[Michael Dalgleish Narrative]: My earliest recollections were as a child. I had robot toys and my mother started employing them in her house work, and by the housekeeper. When I was a teenager, I began to see them more and more all around me in some form or another, as they began to proliferate. I found them fascinating when I encountered them in my everyday world. It wasn't until I was about to graduate college and toured my father's business; I worked there as an intern, and thats when I really made the connection between him and them. By the time I was a young independent adult, I believe he had some idea what he had helped to unleash, but by then, the course was the course, it was set, and he was not likely to turn it back from there. I believe he thought he could alter the course along the way.

Michael recalls faint and foggy images of a cleaning machine crawling across the floor in his childhood home. He sees himself as a child, 8 years of age. He is in bed at night, and his father, Boirix, is reading one of his two favorite childhood stories. One is titled, 'The Ant and the Grasshopper', and the other, 'The Little Red Hen.' Both tell moral stories for children that instill the value of hard work and personal responsibility. Then, he is watching and playing with the robotic cleaning machine like a game, or toy as it attempts to do its duty.

He recounts his time as a youth in school as the foggy images in his head move as if progressing through about 7 years time to a teenage Michael in high school. He relates a story he remembers when walking down his high-school hallway, where he notices a small robotic 'cyber-scooter' machine consisting of a motorized electric scooter with wheels, gyroscope, and handle bars. It has a tablet computer mounted on it's front to display a face with the tablet's camera for eyes. It is remotely piloted by someone, while it follows him around about 20 feet behind. It dodges by turning away when he looks at it. At first he pays little attention, but a few days later he sees the unit standing in a huddle with a group of teenage girls that giggle and stare at him as he walks by. Later on he is curious, so when he sees the unit driving itself down the hallway looking for him, he hides from it. He sneaks up from behind and jumps on it and drives it off into an empty classroom, determined to find out who it is 'cyborg-stalking' him. He flips the tablet around and finds a young girl covering her face, too embarrassed to reveal who she is. He eventually coaxes her out and meets Meghan Davidson, who has had a secret crush on him for about 6 months. Meghan is recovering at home from an automobile accident that left her unable to walk without considerable pain, so she undergoes daily rehabilitation. They become friends and sweethearts, and for the next several weeks, Michael is seen riding around on the scooter machine with the screen flipped around backward while he talks to her.

The unit was put together by the schools maintenance crew, by attaching a tablet computer and running a cable between both machines ports. It is remotely operated by a joystick from Meghan's house where she is confined.

Success is often copied, so after the romance is sparked there is a rash of other little girls faking a stay at home illness, demanding to be cyber-schooled, secretly hoping to find a 'Robomance', as they call it, by cyborg-stalking their hopeful flames. The school is required to comply by law, so there are nearly 15 such machines with tablet faces seen buzzing about daily before the end of the school year. Michael himself fakes an illness for a week to get himself a cyber-scooter, so he and the Meghan-bot are seen together with tablets face to face.

The images of Michael's memory progress in time again. He remembers a quick flash around the time of his college graduation where he is walking down a corridor in an office setting with many busy people bustling in and around. He is walking with his father Boirix Dalgleish. A tall man wearing a white lab coat passes by the two and gestures a greeting. Boirix spins around facing toward the man walking away and calls out, "Oh Jim, I would like you to meet my son Michael."

Jim, James Bowen, then stops and spins around, and the 2 approach each other.

Boirix announces, "Michael this is Doctor James Bowen, the company's director of research!"

Michael reaching out his hand to shake the other mans hand, respectfully exclaims, "Hi I'm Michael."

James Bowen, with a smile and a little bit of a jesting smirk, jokingly says, "Hello Michael, nice to meet you, you look like your father, sorry about that!."

Boirix then says to Dr. Bowen, "Jim come see me later, I'd like to discuss how the new system is coming. I just had a few small issues to go over."

§

Michael continues reciting his story to the eager audience at the podium.

[MD Narrative]: I was born in Texas in 2005, and attended high school and college there. I studied a number of subjects including computer-science and engineering, with a minor degree in economics, and some courses in political science. In many ways though, I was always a student of these things. One of my first and proudest achievements as a boy was to win an award in high school for being a young entrepreneur. That event reinforced what my father had already instilled in me, and began to shape my attitudes, and how I came to view the world. My attitudes toward work, money, politics, business, and issues like taxes, and the compulsory redistribution of wealth. These issues were really hot topics when I was a young man; around the time of my high school, and college graduation, there was much fighting amongst the population of the country about them. The country was falling apart, and total failure was imminent, stemming from the considerable burden of debt that had accumulated from misguided government policies put into place over a number of decades.

I was very opinionated, angry, and outspoken about many issues, and my attitudes and beliefs really reflected those of Boirix, my father. By the time I started college, there had been a major crisis in the country; the government had virtually collapsed due to the weight of the debt and ill-conceived trade policies that over time, had allowed the erosion of our industrial base to the 3rd world, which nearly caused the total collapse of industry and the entire economy of the nation. It was resurrected by the people finally taking a stand to stop and reverse the destructive *policies that favored the interests of global elitists over the nation as a whole. My father was among those that really spoke out and demonstrated a clear vision to reverse the path; to begin to restore and rebuild our nation.*

Desperate Acts of a Desperate Nation

[MD Narrative]: The executive and legislative branches of the US government set everything in motion in the early 21st century by passing an act that launched the course, and like most any ill conceived idea issuing forth from these less than honorable bodies, this one had numerous vast unforeseen and unintended negative consequences.

Having averted total disaster, the act had laid the course for the destruction of the very body that set it in motion and much more far reaching fallout for the rest of the world. It was a crossroads point of fate in the history of civilization. Rest assured had the act not commenced from the US government, the course of history would not have been altered significantly but the same inevitability would have commenced regardless. The course is the destiny of man, it is our fate and follows from our nature, and the nature of the world in which we live. It is the inevitable course.

Productivity Theory

[MD Narrative]: The impetus for much of the move toward automation was economic, reducing costs in order for the producers in the developed world to compete with labor from the emerging world, and to restore the economies of the west.

The profit motive was paramount. For centuries merchants and producers alike realized the competitive advantage of cutting material and in particular labor costs, and the profit potential that could be achieved by small incremental advances in machinery and technology, which resulted in significant advances in productivity, and thus profit. Many highly regarded business schools taught graduate level courses in the subject, and there was no shortage of eager students.

Moving Production to the Labor

[MD Narrative]: The politics of the move of production to the cheap labor was driven by the profit potential, and the rapid expansion of the massive available labor pool abroad, which in the previous decades was limited to those educated and

available in the industrialized world, or those living in close proximity to the location where materials were processed. After the migration of industry, that same factory which previously operated exclusively in the developed world, was now in close proximity to literally billions of newly available able workers, so the jobs of the developed world soon followed. If one producer moved operations, all his competitors were forced to move as well, or face shutdown, but they all retained access to the gigantic marketplaces in the developed world. This all ignored some fundamental principles of economics, which can be summed up as: 'Create jobs, build the market! Eliminate jobs, lose the market!'

These workers were able to provide labor at less than 5% of their industrialized counterparts. But what about the cost of materials? Materials costs are largely determined by the cost of labor as well. Low labor costs, combined with greater access to new technology and productive tools, all in close proximity to the raw source of materials, significantly reduced the cost of materials. Now overall, a manufactured product, and many services became much cheaper to produce.

The impact of the industrial migration was that the costs of manufactured goods and many services were reduced by many orders of magnitude over what the same costs had been before this era, thus the retail price of manufactured goods eventually dropped by up to 60% over the previous era.

Americans began to find products at a significantly reduced price, but could not compete to produce them on labor costs, and the jobs which remained had to become competitive in some measure, or they were eliminated. The advantage of the arrangement was increasingly obtained at the expense of their jobs and lifestyle, while significant profit accrued to the foreign producers, and the common people of the developing world, who could never have had access to those things before, but now found them affordable, and, for the first time, they had the jobs and money to buy them. On the surface, it seemed equitable, initially, but would soon yield considerable trouble at home.

Prosperity

[MD Narrative]: Consider an interim condition, a glimpse of that which happened during a transitional period of the 21st century. Consider that according to economic theory as once practiced, the greater the productivity in any economy, the

greater the collective prosperity of the population. *Productivity can be defined as the ratio of things produced to the cost of the production. The cheaper products are made or services rendered, the higher the productivity, and the more accessible they are. The greater the increase in productivity, whether a factory, accounting firm, or farm, the more there is produced, and that which was attainable only by a few before, is now attainable by the many. What was once only had by the rich can now be had by even the common, who have become more like the rich as a result.*

Economic Dissipation

[MD Narrative]: In the early 21st century, between the years of 1960, and about 2020, the rapid move of manufacturing and other productive resources from the industrialized nations to the developing nations occurred at an unprecedented rate. Up to that time, America was the greatest economic force ever seen, and was driving the world economy. Whether through hubris, or abject ignorance, massive manufacturing resources were actively migrated from the United States and Europe, to Japan, China, Indonesia, India, South America, and eventually most of the world.

Why such shortsighted policies, credited with the devastation of the American economy were enacted is not fully understood, but theories abound. Global integration induces global control. Those having seized the levers of power then, just as before, view the world as their own possession, to arrange and order as they see fit. That seems the most accepted view, and even today, we recognize it to be among the many reasons for such reckless policies.

It was craftily sold to the public as something virtuous, praying upon their naivety and misguided sympathies. Platitudes and hackneyed moralizing about increasing the general prosperity of the worlds impoverished, and achieving some sort of fairness, or parity, in the division of resources and wealth, shared among the populations of the earth. In reality, it was simply a divide and conquer strategy of control, designed to separate the advanced civilizations of the industrialized west from their wealth and power.

Between the years of 1974 and 2024, what was then the modern world and the massive economies created by the producers and markets that were a result of industrialization by less than 10% of the worlds population in the previous 250 years, were dwarfed by the migration of those productive resources to another 75% of the world and eventually the entire world. Initially, the economies of the developed countries exploded, and the economies of the poorer countries became very good, as what were once, very expensive productive processes, suddenly became less than one third of their costs previous to that period, but eventually the bloom went off the rose, exposing the failure.

Massive Growth

[MD Narrative]: It is easy to understand that with massive capital infusion and resources from the developed nations into the underdeveloped nations, combined with their access to an already robust and burgeoning marketplace in the US and Europe, which had been developed during the previous 250 years, where trillions of dollars from consumers was available to absorb this new production, massive economic growth ensued, and many of the new producers became very rich, and their nations grew at unprecedented rates. While, in the industrialized world, the growth from such folly eventually slowed and reversed course, leaving massive obligation, threatening collapse.

There were many steps, many tools, and much political maneuvering employed in the great productivity migration that transpired and the result was a new world in the sense of the elimination of the intensity of poverty in some parts of the world, but significant dislocation of employment and security in the more developed world. Initially, the sense of increasing prosperity seemed to be universal by all participants, but soon faded after having produced the desired effect of those that engineered the circumstances.

Rosa the Housekeeper

Standing in the kitchen of their home in Dallas, is the new housekeeper Rosa Cárdenas, recently hired by the Dalgleishs'. She is an illegal alien from Mexico, although the Dalgleishs' are not aware of her illicit status, due particularly to the fact that she came highly recommended from an agency, and in possession of what seemed to be legitimate identification documents, and with glowing, almost ecstatic references from other families she had been previously employed by. Although brand new to the family, Celia Dalgleish admires her warmly and relates to her now as if she is a trusted member of the household that includes the Senior Dalgleishs', and their younger children, and on occasion, Michael.

[MD narrative]: I remember my first meeting with her, and my distinct feeling

that there was something that I did not trust about her, but while applying for the job, according to my mother, she seemed to have a very calming, and uplifting, almost hypnotic feeling about her that made people feel at ease and trusting. It was only about a month after she was hired, while I was at their house on a warm saturday morning, shortly after they got up; I was sitting there with mom drinking a cup of coffee, talking with her, when she walked into the kitchen.

"Hallo misa dog-leash, Hallo meester dog-leash, how joo today?", Rosa greets them, as she stands in front of them with hands clasped together in front.

"Rosa, we've told you, its doll..., gleesh ", Boirix replies, while looking up at her from his terminal device.

"Oh-kay meester dolg, leesh..., I know it how is said", she reassuringly replied, with a taint of disrespect in her voice, and a little attitude as if to say, "ok, just keeding, keep joo pants on big boy."

"Michael, I would like to introduce someone to you, this is Rosa..., and this is Michael our oldest. She is helping me around the house now; she came highly recommended, didn't you Rosa? Wasn't it the uh, Jensen family you were with, for... , was it, 5 years, taking care of them?", Celia asks as she flips through papers sitting on her lap.

"Jyase, fy yeers I work for Heenson fally, then they no want me, but is OK now! Now Rosa gah new dog-leash fally to work", Rosa whines in a nasally voice.

As soon as Rosa exits the room, Celia Dalgleish says, "Well she did come highly recommended, but strangely, her references said she spoke perfect english, but she needs to work on that somewhat."

Boirix replies, "Yeah, if you can believe the documents are real, not faked."

Then Celia says, "Bo, how can you say that, do you really think she is capable of faking something like that, and why would she need to, she seems very sweet the way she is, who wouldn't want to hire her, and besides she comes really cheap; I'd like to hire several more like her!"

Michael says, "Mom, they fake documents all the time to be able to work here, and they have gotten quite good at the fakes."

"Yeah, right about that son, we're being overrun by them!", Boirix declares.

[MD narrative]: Little did I know at the time, but they would pay a terrible

price, by letting that woman into their home. My youngest children may have become victims as well. I thank God to this day, that I did not take them over the night before that day, like we were supposed to.

CHAPTER TWO
Unslumbering

Knowledge

In the beginning God created the heavens and the earth...

...And God said, Let us make man in our image, after our likeness: and let them have dominion over the fish of the sea, and over the fowl of the air, and over the cattle, and over all the earth, and over every creeping thing that creepeth upon the earth...

...And the Lord God formed man of the dust of the ground, and breathed into his nostrils the breath of life; and man became a living soul...

God created a man Adam, and set in the earth, the land of Eden, and in Eden he planted a garden. In the garden no entropy is found.

...And the Lord God planted a garden eastward in Eden; and there he put the man whom he had formed...

...And out of the ground made the Lord God to grow every tree that is pleasant to the sight, and good for food; the tree of life also in the midst of the garden, and the tree of knowledge of good and evil...

...And the Lord God took the man, and put him into the garden of Eden to dress it and to keep it...

...And the Lord God commanded the man, saying, Of every tree of the garden thou mayest freely eat:

But of the tree of the knowledge of good and evil, thou shalt not eat of it: for in the day that thou eatest thereof thou shalt surely die.

Economic Entropy

[MD Narrative]: The increasing export of the industrial base of the country over the previous few decades was only part of what had precipitated a crisis situation, but was not the only factor; there were other causes in the cascading chain of failures that ultimately befell the nation. The crisis was composed of significantly diminishing growth of the nations economy, which only contributed to the other factor of unmanageable debt, which had accumulated over decades and began to accelerate as the government continued to spend money it borrowed from others, based on flawed leftist theories of economic growth, through government busywork programs, in an attempt to keep the house of cards from collapsing in on itself. The flawed theories inevitably proved fatal, and were thereafter widely discredited, but a real solution was still needed, and fast.

In around 2017, the government finally started to reckon the dire situation they were in vis-à-vis the escalating national debt, and the obligation for numerous social welfare programs. Programs like social scrutiny, Mediocare, Mediocaid, and a host of social welfare programs, burdened by the complex of health related expenses required to support an aging, and increasingly non-functional population. Then adding to the pile, there were massive amounts owed to foreign governments, foreign and domestic entities, and citizens of the country, who demanded payment of the debt with interest. The people never really understood the magnitude of the debt, nor how to deal with it until it was too late, due to political considerations, and because the truth about it was manipulated by authorities, with the help of a complicit media, fearing loss of their own position, privilege, and perks. Although as unmanageable as the accumulated debt was, it was not real in that the same process that had created the debt, had also rendered the currency increasingly worthless, and consequently, the debt is only as good as the currency in which it is denominated. Bye and bye, the US government had written a check that the American people's asses could not cash, and collapse was inevitable.

§

America was the greatest economic engine the world has seen up to then, but like any engine, it could only carry a limited load before it started to sputter and eventually fail. That is what happened, and along the way we began to take on the characteristics of emerging third world countries, while many of them began to look like we did only a few decades before; we were so prosperous then, but it all started going bad. We became victims of our own success, and hubris. The current political class of dubious worth, recklessly mortgaging the fortunes of future generations, profligate well beyond our means, believing the party would last forever; reality finally caught up with the nation. Once the deterioration gains momentum, its hard to reverse direction, or gain consensus to move forward again while the increasing cascade of failures begin to accumulate. Proposed solutions abound, with virulent, and monumental resistance to most, while individuals and advocacy groups only care about their particular situation, none willing to sacrifice for the good as a whole any more than one pig will share the food in front of it with another. So it was with those generations!

The American Republic had degenerated to resemble what was described by the 19th century political scholar and philosopher Alexis de Tocqueville, in a bygone era when he noted:

"The American Republic will endure until the day Congress discovers that it can bribe the public with the public's money." … and

"A democracy cannot exist as a permanent form of government. It can only exist until the voters discover that they can vote themselves largesse from the public treasury. From that moment on, the majority always votes for the candidates promising the most benefits from the public treasury with the result that a democracy always collapses over loose fiscal policy. The average age of the world's greatest civilizations has been 200 years."

The country had certainly taken on the characteristics of much of what was described by de Tocqueville, and it had been slightly over 220 years from the founding of the republic, so true to the prediction, imminent total failure was bearing down fast. The difference in this case was that the US had become such a large part of the worlds economy, and the collapse was predicted to create such a tremendous disruption to the world, that, out of nothing more than pure desperation and not knowing any better how to avert the impending disaster, it was decided by the sages in the nation's capital to make attempts. The house was thrown wide open for any and all manner of absurd proposals to be heard and considered; the philosophy of: *"Throw enough mud at the wall, and some will stick. Lets see what sticks."*

It required extraordinary leadership, and there were hard choices to be made balancing the limited options between making good on the social obligations in a meaningful way, and maintaining economic

17

solvency, while cultivating vitality in production and growth. Most of the solutions proposed were politically impossible, and not tenable by most. The floor of the people's house was thrown wide open for hearing, with all manner of nut-jobbery and crack-pottery proposals entertained, as their advocates started to crawl out of the woodwork.

Hearings on Capital Hill in Washington DC commenced in the early spring of 2018, soon after several large banks shut their doors, and the markets plunged by 65% in 3 weeks, with the next plunge expected within a few months. The bond market became very skittish with investors around the world nervous about buying anymore of the paper sold by Washington, especially since Asian and European investors bailed on several recent bond auctions. There was simply no interest, so no one participated, and interests rates shot to nearly 15% after having been thought to have stabilized at 10.5% just 2 years earlier. This signaled that the government was not able to borrow anymore money to meet its obligations, which left only the Fed Hoard printing presses to cover the mess. Inflation had already been reported at 9% and heading up from there. The price of gold had topped $3600 per ounce, and oil hit $220 a barrel; a significant amount back then, and both were heading higher quickly.

The house is in session and Representative Henry Stumble is holding hearings on proposals for ways of arresting and reversing the financial slide. These hearings have been going on for the better part of 2 months, and are in session for 10 hours per day, while there is very little other business being considered by either houses of congress. Both chambers are holding hearings of the same nature, with the house enlisting more private organizations and groups, and the senate hosting hearing sessions from business and industry.

A group is sitting at a table in front of a large committee, which are nearly all present, and includes hundreds of other members from the entire house. The 3 witnesses are testifying about their proposals to sell or lease several southern states to Asian governments and sovereign funds, who will effectively own and run the states. While they occupy the land, they would have control over all the affairs of the state, enabling them to collect rent and real estate taxes from the citizens, along with tolls from infrastructure assets, and fees from businesses based in the region. They would absorb all of the social costs, then split the excess booty with the government in Washington, thus reducing government overhead and increasing revenue.

"We think the less valuable southern states, like Alabama, Mississippi, and Louisiana would be preferred over any of the more valuable ones in the north of the country, or even the upper-midwest, for several reasons, mostly having to do with the fact that asians already do business, and manufacturing in several of these states..., uh", he is cut off by a question from representative Jesperson from Kansas.

"Excuse me sir.., uh, Mister Gupta, how do you propose that the revenue is split between the two? Also, if you will sir, ummm, have you considered how the assets are be delivered..., in other words, would we just turn-over the statehouse and enforce or sanction police action on the citizens, or involve the governor, or would it be more a wholesale delivery...?", he is cut off by Representative Stevens from Louisiana, "Please gentlemen, lets get some balance here, we are not going to auction states off, no matter how much is owed to foreign governments. They can threaten and demand eminent domain all they want, it is simply not going to happen..., ever, unless, well, I'm sure the gentleman from Kansas could always volunteer his state be the first!"

The chairman then dismisses the witnesses and announces the next group to enter. He asks them to state their names for the record. The two swarthy and disheveled men are iraqi brothers. They and one woman take their oaths and sit down, and begin their testimony.

There is bustling on the podium and the upper gallery as the staff passes papers that contain the language of the proposal out to member representatives present. A few minutes pass as they look at the material, then Representative Stumble states, "Gentlemen, ma'am, are we to understand that you are here to explain to us about a heroin cultivation and distribution system...?", the witness clears his throat, then speaks up interrupting him, "Yes, senator, thats right, we think US government can make..., uh, increase money to the treasure by....!"

Representative Stumble then interrupts him, "Its representative, not senator, this is the house chamber, not the senate, we are all representatives here, I'm not a senator..., not yet anyway..., I'm sorry, please proceed."

The witness then continues, "like I was saying, there is a lot of money in heroin, and if heroin is legal, well, my cousin in Islamabad has a large farm. He can supply...!"

He is interrupted again by another congressmen, "Mister almoo-sawi is that right..., yes good, mister moo-sawi, you are aware that heroin is illegal in this country, and you are proposing that the United States government deal heroin to its citizens. Is that right?"

Responding, the witness continues, "Well, no sir mister senator, we would only guarantee the supply. You could sell it however you think will work best."

Representative Stumble then crashes a gavel down, and closes the discussion, "Ok, gentlemen, we need to move this along, thank you.", ending the session.

Social Scrutiny

[MD Narrative]: The social safety net programs that had been sold to the population by political elitists, were popular due to the idea of getting a so called 'free-lunch'; they were euphemistically referred to as the 'Free Money' programs, something that in that era was said to be impossible, as in 'there is no free lunch', yet the majority of the population believed they were personally exempt from the universal laws that describe, and govern equitable arrangements; they believed they were the exception to the rule for whatever reason justified the self-deception. The social scrutiny system was described, and believed to be a kind of government managed insurance policy against old age, which was ridiculous because insurance was always based on actuarial probability and risk, and there was never any risk or chance in the question of growing old. Even today we recognize the fallacious manner of such a faulty belief. Greed, being a more powerful motivating force than the restraint and sense which age and experience bring, made it easy for people to believe the lies, and ignore what they already knew to be true, because of how it was sold to them, exempting them from the responsibility of diligently vetting such a devil's bargain.

Earners simply allowed the money taken from their wages, and paid into the government managed fund, and trusted that they would be taken care of at some future point in their lives. The government accurately and often reported the activities regarding the money paid into the fund, yet people chose to ignore the fact that shortly after the money was paid-in, it was immediately diverted and wasted elsewhere, and what replaced the stolen money, was a series of worthless government bonds. They were worthless because nobody ever expected the government would be able to make good on the mounting obligations, so like all pyramid schemes, a stream of benefits were paid as long as the government remained solvent, then it collapsed.   Throughout most of the history of the program, eligible recipients received benefits way in excess of their particular contributions, thus putting the fund into continuous deficit, and requiring the new contributions to remain in operation.   When the younger earners began to realize this fact, and the inevitability of insolvency, which meant they would never see a penny of benefit, they opted-out massively.

§

The house is in session with congressman Henry Stumble residing. The hearings are a continuation of the discussion for finding a solution to the impending fiscal collapse, but now the testimony has turned to proposals for considerable wholesale draconian cuts from the budget, including cold-turkey stopping, or massive reduction in social scrutiny payments, reasoning that the system has been in deficit for virtually the entirety of its life, and the money is paid to people that contribute very little in the way of productive measures, while those on the other side of the divide arguing to the contrary, the record payments currently being made to seniors are a vital stimulus to the economy, and required to avert an accelerating decline in the economic output of the nation.

Testifying is the government bureaucrat that currently heads the Social Scrutiny trust fund for the government. He is testifying on solvency issues. Congressmen Stumble asks him to proceed with his statement, "Mister Ponsey, could you address the reasons, as you have stated, why you believe the fund is still solvent, and maybe give us an estimate for how long, or when you believe it will become insolvent?"

The witness, Carl Ponsey, with the superb double-speak that only an extremely skilled politician can muster states, "At some point near the end, which we believe to be very soon, the problems will intensify and are already starting to escalate…, uh…, mainly due to the increasing worker to retiree ratio, and the diminishing workforce, meaning fewer workers paying in, and many more drawing benefits…, so the entire system is threatened. That and the fact that most of what has been paid into the fund over the years has already been diverted, misappropriated, and spent. However, as we stated, we believe the system will remain intact and operating fully funded well into the future. We believe there are sufficient funds available now and into the foreseeable future."

Despite double-tounged assurances, the reality was the demise of the Social Scrutiny system was hastened with imminent failure in the short term. The nation as well as most in Europe were suffering the ravages of an unmanageable debt burden, which showed no signs of being brought under control any time soon. The self serving political class feigned concern, but only gave lip service and actually added greatly to the problem, so were duplicitous hypocrites in dealing with the impending crisis.

The session continues with other witnesses giving testimony. A. Finkel, professor of economics from a northeastern college is up next explaining that the problem with the mounting debt and continuing fiscal collapse is really because according to current accepted, rational economic theory, the taxpayers have not reckoned their part, meaning the government has not done enough to stimulate the economy. Although not his words, the meaning is clear; as far as he is concerned, there has not been nearly enough money wasted by throwing it down a rat hole, handing it out to friendly causes, massively spending on worthless and destructive social engineering, busywork, foreign adventurism, and a host of other projects which are favored by the professor, and virtually all of his elitist ilk. This of course is something that glaringly flies in the face of the facts, and is not outside of expectation for people of his mental and intellectual disposition, having made recommendations for decades advocating incredible amounts of public spending on dubious programs.

Speaking, professor Finkel lays out his premise, "In our estimation, additional trillions of directed stimulus are needed for the next 3 to 5 years to drive the economy into a mode for sufficient growth necessary to avert fiscal disaster."

As he is speaking, a grandstanding congressman, named Schuman, looking to go on record with a witty quip stops him and asks, "Professor Finkel, may I ask, and please correct me if I am wrong sir, but are you saying we need more of the hair of the dog that chewed our legs off? If you will indulge me sir, we have heard that for years, and it has only gotten us into deeper water."

The professor answers, "You may choose to characterize it that way congressman, but the evidence is clear and unquestionable. This has been demonstrated over the last several decades. The consensus is unanimous amongst the vast body of economists of any credentials. Those that say otherwise are either not particularly well informed, or have an agenda contrary to what is of the best interests of the nation. Again, I stress, there is absolutely no sun-light between those that have always been authoritative on these kinds of issues. It is my fear, that the course has been altered too dramatically away from the correct course already, and it may be too late to correct."

The civilian audience present in the chamber sneer and loudly make raucous and insulting comments indicating their contempt and disgust with the comments from the mouth of the professor. Even some of the congressmen present, including both congressmen Stumble, and Schuman, sensing the mood of the crowd, join in and call for the man's comments to be expunged from the record, remarking how that thinking has run its course and is now becoming outdated.

§

A large crowd of people protesting outside of the capital building are growing more and more agitated. There has always been temporary groups of people demonstrating their disdain with the way things are going for years, but now the crowds are permanent. They are squatting and defying orders from police to move out or face arrest, and the local jails cannot possibly hold their numbers. The promenade has begun to resemble a refugee camp, with tents and trash, and the ever present groups there with their pickets signs, platforms, bullhorns and loaded rhetoric. Most of these people have been unemployed for years, or are less than sufficiently employed, and while waiting for work, they make the promenade in the nations capital their home. At any one time there are nearly 25,000 people present and the omnipresent political tension hangs in the air like a thick and foul smog.

Increasingly, on recent occasions, the area gained a new presence with loads of elderly people having made the same trek, who started to become just as permanent a fixture and part of the backdrop as the younger folk. Their motor-homes and assorted recreational vehicles parked in, on, and all around, any and all open grassy areas, the parking lots of government buildings, along the avenues which parallel the promenade and across the river all around the pentagram building, and even making the monument to the obelisk a parking-lot and commando headquarters for the ranks of their organizations guerrilla operations. They have come to voice their displeasure about the fact that their stipends are diminishing with respect to already ravaging interest rates, and the accompanying inflation. Food, energy, and other basic costs are skyrocketing, but for the last 28 months, their social scrutiny allotments have not increased, and in many cases the amounts have been arbitrarily reduced without explanation. Some of them have found themselves informed that they have been arbitrarily classified as deceased without appeal, removing them from the roster of beneficiaries, so their checks have stopped altogether. Many having no ability to hire an attorney, and with no avenue for recourse, all the while finding their letters and demands falling on deaf ears, they have now decided that there will be hell to pay and are going to make their presence known, and their voices heard loud and clear. Enmasse they cash-in, or liquidate their nest eggs and head to the nations capital.

Once there, the difficulty is that the promenade is already occupied with those on the other side of the divide that are saying something quite the opposite, and they are not willing to give any quarter to these upstart usurpers. They are the young, able-bodied, and working-class and have already declared war on being personally taxed for the benefit of one group of people at their expense, regardless of the fact that they are elderly, some are family members, and none of them nearly as able-bodied. The fact is, the social safety system has now been revealed to the young as a scam, a Ponzi scheme of the first order, and there are considerably more of their ranks willing to throw down with any of the others present, agitating, making noise, or advocating for restoration or continuation of the scheme. They have declared themselves perpetually exempt from paying into a system in which they will never benefit, while they are chided by the elderly recipients that lecture them on their civic responsibility to work and pay taxes to insure the elderly, weak and vulnerable are well taken care of.

These elderly warriors have gradually managed to finagle and carve out a staging area on the monument where they have set-up. With several large groups of recreational vehicle battle wagons parked in and around the promenade in grid formation, and canopies out; they are dug-in and girded for battle. They have set up a large covered structure that can seat nearly 100 people, with chairs and tables setup, and meetings in continuous progress.   They are amply more organized than their younger foes.   Several stations are set-up where little old ladies are busy hand making picket signs with various slogans; or war cries may be a more apt description, while handing them out to a line of protesters prepared for the battle.   *"My Pension My Right"*; *'I Paid It, I Intend To Collect It"*; *"We Are The Generation - Don't Trifle With Us"*; *"Paid Retirement Is A Human Right, Earned Retirement Is An Enforceable Contract"*; *"Diminish Social Scrutiny At Your Peril"*; *"Don't Mess With The Blue Brigade"*; *"Eighty Today, Ninety Tomorrow, And I May Outlive You"*; *"I May Look Dead, But I Still Vote"*, and so on.

These folks are determined to do whatever it takes, so they have had to resort to parking the bulk of their rigs in various locations in and around the city, but have found ways to daily transport their ranks from near and far, some using golf carts, which they also use to truck their masses to the battle front, for a strong showing of numbers whenever the occasion demands it.   In a small area outside of the main protest, a line consisting of 10 to 12 seniors are faced off on a grassy area carrying picket signs, and vociferously proclaiming their non-negotiable demands, and making their case, *"We have made many sacrifices to make this nation as great as it has ever been; we gave our lives in world war two, and Vietnam, so you could be free to work, and raise your families."*

"We sacrificed to give you the ease of life you enjoy now; don't deny us with the little time we have left, you will be where we are soon enough."; *"Our children owe us the dignity of living the balance of our lives independent of burdening the independence of our children!"*, and so on.   Some of the more animated seniors were more raucous loudly proclaiming slogans and chants with a bull horn, reminiscent of the Vietnam war protests, *"Hell no we won't go!"*; *"Hey hey uncle sam, why would you kill your gram"*; *"two three four five, we are very much alive"*; *"Decide, justice not Gericide"*;

The weather has been somewhat cold in the late spring of the year, so it has been raining heavily for several weeks, creating a not very pleasant surrounding for the participants braving the elements, who put up with it as part of the price required in the effort to carry their concerns to the nation.   It is cold and damp, and there is considerable mud in and around the ordinarily grassy areas of the promenade where the younger tented occupants reside, who have been tromping everything underfoot for the better part of the last year and a half.

A group of elderly people are in this environment facing a larger group of 25 to 30 of the younger protesters. Many rival groups of the national news media are there reporting as they have been for the last several months, keeping an eye on the situation, and some have recently began to get up close, doing interviews, expanding with some detail on the divergent views of the factions, endeavoring to pinpoint the heart of the controversy. They have taken up camera positions and backdrops that frame the stand-off. It is supposed that some of the protesters on both sides played for the camera, becoming more animated and agitated than they ordinarily would have been otherwise, because the confrontation that ensued was extraordinarily vicious and more brutal than anything seen prior, and that episode really started to shape the standoff and galvanize the nation to finally make a change about the issue.

A group of about 5 young men decided to get in the faces of several old blue-haired grandmothers encroaching on their turf, and shout vicious and nasty things at them, attempting to move them along, but the grannies held their ground, walking about with the pickets in a circle and forming a Maginot line of defiance and defense. In some cases, the aggressors were just trying to get some sleep and were tired of the droning-on by the elderly protestors. The young men told them to move on, but when they would not budge, they attempted to intimidate the grannies by lunging at them and taunting them, but the elderly stalwarts were not moved and instead chided them telling them to learn bathing and personal hygiene. Finally, one of the men tackled an elderly woman knocking her into the mud, simultaneously, another man dropped down on his hands and spun his legs around martial arts style, kicking the legs out from under the husband who had come to the aid of his muddy fallen wife. He too landed in the mire, and the two of them splashed around slipping and struggling to extricate themselves from the muck.

The whole thing was captured on camera and run repeatedly in the news with much commentary and condemnation for weeks, ostensibly to garner sympathy for these elderly and their cause, but quite astonishingly, it actually worked in reverse and opinion instead coalesced in the opposite direction, with some even cheering the action of the youthful brutes, while the elderly participants were quite possibly for the first time in the nations history, depicted and perceived as high-minded and pampered mobile pariahs and blood suckers, sucking the health and vitality out of the nation, demanding they be maintained while families and the young were struggling nary able to put food on the table.

This event helped turn the tide to some extent with public opinion concerning most left wing ideology and anything resembling government arbitration in the lives of the average person.   These programs were scrapped wholesale, with many incidents of those that advocated these programs, or the growth of government being challenged, shouted down, or outright literally chased out of town, or given and old fashioned street thrashing. The issues concerning social scrutiny and its myriad companion programs like mediocare, mediocaid, socialized medical care of any sort and welfare, were at one time considered to be the so called 3rd rail of politics, meaning that anyone advocating cutting or even modifying the programs were likely to be voted out of office mostly by these old guards. That power had now been broken, and politicians were grateful that they were now free again to do more of what is expeditious and advantageous for the nation, rather than pander and give-in to power groups, regardless of the consequence. It then became popular to denounce and take pleasure in proudly announcing scrapping the wasteful programs. Successive political campaigns came down to a contest of claiming which politician had scrapped more programs than his opponent.

§

[MD Narrative]: A defective stop-gap solution had been proposed all along, which was to increase the age for qualification of benefits by retirees, and increase taxes to fund government obligations, but this simply failed, because it would never produce what was necessary, so it led to violent protests by both sides of the divide, and created an incentive for people entering the work force to opt-out of an ill-conceived system with rapidly accelerating losses, in favor of more reliable, self-funded retirement plans. The last social scrutiny payment was made to a recipient, and the program was halted altogether sometime in 2023.

The Sub Standard Labor Solution

[MD Narrative]: Some voices proposed massive increases in immigration and cheap labor from the 3rd world to offset the problem, but this was shown to be a political hot potato, and would have little effect in solving the problem, and in fact it only

negatively impacted the problem because of the social costs of care for these people that could never earn enough for their own keep. The benefit of their cheap labor only accrued to large corporate interests, who demanded their entry, but using their political clout, managed to pass the costs for the burden of their upkeep on to the public. Likewise politicians also advocated for the massive influx of 3rd world labor, wanting them as voting constituents, the consumers of government planning and programs, and to build the ranks of advocacy groups composed of their own ethnicity, needing their numbers for the purposes of lobbying for the power that extends from special government consideration and money.

§

The American people have become increasingly tired of trade policies that favor all the nations with which America trades, at the expense of American industries and the American worker. Mexico is one of the greatest recipients of favored trading status, and as such has been shipping billions of dollars of manufactured and value added goods across that border for the better part of 3 decades, all the while, American jobs have been displaced at a proportional and steady rate. Now the tide has started to turn, and the goods are not being accepted at the border any longer. A several million strong, multi-state posse organization has seen to that. They are the result of a political backlash against immigration, and a stand off between the government and all their intrenched interests, an increasingly aggravated populace, tired of being kicked in the teeth at every turn by elitists buying the process; who have access anytime they want to those at the helm of the federal and several state governments.

Trucks are being stopped and turned back by force of arms, or these men are simply shutting the border crossing stations down. Americans have had enough of the political status quo, and have started to take charge of their own destiny, instead of letting the political elitists and their lap-dog minions in government make all the decisions. There has been a wake up call, and the 'Average Jim' is getting into the act and realizing that his future, and the future of the nation requires his direct involvement. They organize in groups, setup patrols, shutdown avenues and arteries of ingress, develop and utilize various means to deter the incursion, and create ad hoc policy directives effecting the entire region along mainly the southern border. They have been successful in capturing and detaining massive numbers of invaders, which they send back across the border without the invitation, nor intervention of any government agency other than the local sheriff, who is invariably a proud member of the posse.

The recent change has come about for several reasons, among them are the fact that with a looming fiscal and financial crisis, neither state nor federal government has the resources to operate effectively in many areas of their express responsibility, therefore the people are stepping up, taking charge. They have decidedly voiced an opposition to the destructive trade policies that always promised more jobs and greater prosperity for America, but only enriched the middle-men, the upper ranks of large corporations and the political elite, with little or no benefit filtering down to the ranks of the average American. The way they look at it, trade policy, like virtually every other policy, is something that greatly effects the lives and fortunes of average people, therefore they have a right in its definition and implementation, and it cannot be left solely to politicians, nor the intrenched interests that grow fat off of the predatory nature of trade as has been practiced for the past several decades.

In the recent past, the American market has been a target for every 3rd world producer to penetrate, and plunder by undercutting the American producer in his own backyard, while the American producer has had his hands tied behind his back through bad policy. What took the people and producers of the nation some 250 years to build through thrift, hard work, and the industrial revolution, took less than a generation to plunder by globalists. The American producer, worker, and consumer earned the right to participate in the American marketplace, but the political elitists sold that strictly American birthright to the world, at the behest of globalists, for nothing. There is only a remnant left now, and the people have determined that it will be preserved and rebuilt again, starting now.

As the American Republic failed, and Americans started to really feel the impact, the poor people south of the border began to starve and war amongst themselves. As the American economy faltered, those foreign producers dependent on exporting to the American market began to suffer collapse as a result. This meant the failure of many businesses in those countries, and massive reduction in their employment. Mexico faced this fate after having risen up from a poor, massively corrupt, and unproductive mess just 3 decades before, to become a significant producer of goods, as a result of access to export the bulk of it into the American market; they now faced regression back to their previous status.

The immigration argument raged on both sides, except at this point, the political alignment really became polarized as all political parties, factions, and pro-immigration proponents and interests became lock-step aligned against the will of the populace, who were equally resolute in their opposition. Political considerations came into play when virtually all parties and political sides began to push for complete border removal, or otherwise disregard of the immigration patterns starting to form, in direct insult of the clearly articulated will of the populace.   When enacted, the government refused any monitoring or deterrent along the southern border in order to facilitate the influx of Mexicans into the American landscape, something that the people of the nation had clearly voiced an outrage over, however, the political elitists were bent on accomplishing it as an integral step towards furthering global integration, and hegemony.

Regardless of the American debate, and clearly defined unambiguous stance, there were political forces, and other interests that took matters into their own hands and openly facilitated massive illegal entry from Mexico and south America, with the blessing and help of the interests that wanted them.   As the American political system failed, massive cuts in government spending followed, and so did the money for enforcing borders.   What border deterrents did exist were compromised, some being completely torn down, so that a deluge of illegal immigrants started pouring into the country, bringing the violent terror of drug cartels, which their 3rd world culture had produced, and which had ripped their country apart for decades; these came in right along with the illegal immigrants, setting up operation right in the heartland of America.   They constructed and operated heavily fortified compounds, and managed organizations for drug manufacturing and distribution, corrupting law enforcement with murder and intimidations, and collecting billions of dollars in the process, praying on the American public.   America was week, and broken, and had succumbed to the lowest level in its history, and was in danger of being completely overrun by an invasion from the 3rd world.

This of course led to the mother of all backlashes against the established government by numerous governors and statehouses, which were directly impacted by the invasion.   The states most affected, directly repudiated the federal policies and enforced their own. Citizens took the matter into their own hands and dealt with the problem decidedly, and judiciously, demonstrating to the rest of the nation what fraud had been perpetrated by decades of dereliction; the lies, that there was absolutely no ability to stop the flood of illegal immigration, were finally and decidedly revealed to be a nothing more than a massive fraud, as the people stepped up, and showed otherwise.

§

A sizable posse of sorts is patrolling an area along the river that serves as the border between Texas and Mexico.   It is a loosely amalgamated but massive group of men and women from various areas around south Texas, north Texas, and several other states who have set-up their own defense against a renewed onslaught from south of the border. There have been numerous incursions where desperate people have staged a kind of siege in which they assemble in the 10's of thousands on the south bank of the river, crossing over enmasse, attempting to overwhelm the severely undersized official border patrol resources on the north side. It is reminiscent of a scene out of nature with rampaging animals streaming in seemingly random directions, without purpose. How do you stop a determined herd of wildebeest as they charge headlong across the divide? The men of the posse have come because the incursions all along the border are starting to increase in size as the returnees pile up to try again, along with new ones, so the crowd is cumulative. The Americans have come to stop it because it is taking an expensive toll on their farms, private property, communities, resources, and in their neighborhoods, as these desperate people seek refuge and sustenance.

It is a beautiful sunny late summer morning in south Texas, and flying north for about 7 miles along the river, while looking out of the windshield of a helicopter hovering above the spectacle, it is plainly visible what is about to transpire. Deputy Sheriff Thomas Garza is inspecting the scene and radioing the situation back to the dispatch at the county sheriff's office, "Yea, it is bigger than the situation 3 weeks ago, there is one camp after another.   This thing is stretching the entire visible distance, which Walt tells me is about 10 miles."

This is the 6th such 3 day long staged incursion in the past few months along this dusty stretch north of Laredo.   The deputy continues, "I'd estimate nearly ten thousand are assembled already, and more are coming. This would confirm Buck's accumulation theory. Better send out a call to the other counties, and send a general e-lert to the network, were gonna need the men, uh..., tell them we estimate 2 to 3 days for the cleanup."

Walt the pilot, asks Deputy Garza, "Where is Buck, this looks like it might be bigger than the last one?", the deputy replies, "He decided to go hunting on monday; supposed to get back on sunday, just in time for the clean up."

Most of those coming are from Mexico and Central America, and determined to cross the river, many of them having attempted this crossing several times before, only to be caught just across the border and tossed back within hours. They have brought all manner of small craft from inflatable rubber rafts to those shoddily constructed from old used and nail ridden lumber held together with twine and electrical wiring, and even coffins dug up from local graveyards. Their contents dumped where they are taken, they are quickly converted for use as a floatation craft. Some of the crafts have small ropes attached to the backside so they can quickly be retrieved after they have dumped their human cargo on the other side of the river, ready to take the next load across.

Matching them each and every time is an increasing number of posse members equally determined to stop the incursion and send them back. They are assembled with ropes and hooks, and long poles with sharp tips, used to deflate the rafts before they can touch down on their side, or to overturn the wooden craft, spilling the crossers into the drink. They have several lines of 'catchers' behind them to tackle the slippery ones that get through. There are really no rules to capture these invaders, so the posse members use nets to trap them, rifles and handguns to apprehend them as they approach, and even massive industrial sized doses of mace and pepper spray which they apply liberally using large swiveling truck mounted spray devices that can project a quarter inch stream 50 feet. They have also set up a spiraling line of barbed wire floating right down the middle of the river, and again about 50 feet from the shore, to ensnare the swimmers and boats, and the runners that get past the 3rd line of catchers.

These defenses are setup wherever needed for miles along the entire opposing side staging areas, in the back yards of residential neighborhoods; through parking lots; under bridges, and over surface streets; and stretched right across the sage, tumbleweed, mesquite trees, and cactus of the desert. The posse means business, and they have pulled the gloves off. The posse has employed sharpshooters and snipers set up at various locations along the theatre to take out any shooters from the other side, a technique recently employed by the facilitators of the run on the Mexican side when they shoot at posse members, which is designed to generate panic on the American side as the run commences. In the last month alone, there have been 5 snipers killed on the Mexican side by ex army-ranger posse members, and Texas law enforcement, with 1 fatality and 2 wounded on the American side. This situation certainly constitutes an armed invasion, so should really have been addressed by military assets, but the political establishment has become catatonic in recent years, and are simply afraid to act in any decisive capacity. That, in addition to the fact that several governors of the states have simply said no thank-you to federal meddling in their affairs, and making a bad situation even worse, preferring to handle their own problems locally with local resources.

These confrontations have steadily increased in frequency and by the number of participants on both sides in recent months, and have attracted the attention of numerous groups of civil, and Hispanic, and other ethnic rights advocates, who have travelled down to observe the spectacle, and unsuccessfully attempt to interfere. They have recently brought lawsuits against the posse organizers; the governors of the states involved; the local sheriffs in several counties participating; the mayors of cities along the border; and even numerous private citizens and groups in an attempt to stop the non government sanctioned actions. In numerous of these suits the defendants do not even bother to show up to defend themselves, so they loose or are convicted in absentia, but the convictions do not really have any weight, because there is a general low-level state of anarchy, and a general willingness to disregard any opposition or federal statutes concerning these actions, which is easily justified in the clear failure of the mandated government responsibility for protecting the country from invasion, and because they have near universal support of the people that live in the areas effected, so for the most part, the lawsuits and sanctions are ignored. In many notable cases, even the courts and judges have refused to allow the suits to be brought, or to proceed, realizing where their toast and caviar comes from.

§

Two men went missing at a similar event a month and a half prior to the current event, when 3 members of a civil and ethnic rights organization, Greta Jorgensen, Jacob Blaustein, and Pert Haongasu from the Ethnic Rights Enforcement Coalition or the EREC, traveled in from out of town to observe and interfere with the stop, detain, and return action of the posse, but they themselves were stopped and detained by posse members.

One of the 3 members of the civil rights group is standing on a flat platform trailer attached behind a small car parked along a street that parallels the river and coarse of the stop-and-return about to commence. There is a small group of around 15 protesters in front of where the group's trailer is parked, marching around in a circle, carrying pickets and screaming venomous loaded rhetoric at the top of their lungs toward the posse members there to do a job, as the posse streams across the street and toward a wooded area along the river front.

Greta Jorgensen, a director of EREC, has a bullhorn and is speaking out against the participants, "These people have a right to enter here, and your action preventing them is violating international law, and the civil rights of those people. They are seeking political and economic asylum in our country, and you have no right to interfere with that. The right thing to do is to take in and welcome people in need of our assistance. We have always helped people in distress, that is what america is about."

"You are all cowards, if they can not live amongst us, then none of you deserve to live here either. We should capture you and send you back across the border, or maybe to Europe, because this is not your land. This land was stolen from these people, and they are here to take it back!";... "They wash your clothes, cook your food, and cut your lawn. They care for your children, and pick your farm produce. This country would starve without them. The nation is failing because of this holocaust. You will be held accountable for your actions against innocent people. This is their land!"..., on and on with this kind of loaded rhetoric.

While this transpires, the other 2 members of her group, Jacob Blaustein, and Pert Haongasu, are busy behind the area occupied by the loudmouth pied-piper of propaganda, attempting to block access to a small path that leads to the rivers edge, intending to hinder any posse members from operating in the area. They have already used wire cutters to cut away and remove a segment of barbed wire laid out along the south side of the street intended to snare any runners that get past the group. Jacob Blaustein uses a GPS system, and is radioing the opened location to the incursion facilitators across the river. This is evident as the mass start to lineup in the staging area immediately opposite of the groups position.

Blaustein and Haongasu attempt to stand in the way as several posse members arrive and start to walk down toward the path through a canopy of trees and underbrush. Both Blaustein and Haongasu shove or confront several of the posse members and block them like they are playing a game of football. This angers a few of the posse who have heretofore ignored the group for the most part. A few of them then grab the 2 men and punch them knocking both of them to the ground. They then hogtie both of them, and put tape over their mouths, while one of the posse members goes and gets his pickup truck. He backs it up to the location where the 2 are detained by the group, then load them into the covered back and locks them in. The man then parks it in a back section of the parking area and returns to the event, intending to leave them until the event is over. He then stakes the anchor that holds the leash of his doberman dog immediately outside the vehicle to deter the 2 occupants from any attempt to escape before the melee is completed.

After the event, Greta Jorgensen headed back to the house where she had been staying and remained there for another day and a half, waiting for the men to show up, but they did not. She then made several calls to friends and colleagues inquiring whether the men had reported back yet, and getting no satisfaction, she went to the sheriff's office and filed a missing persons report, then left town. There was an inquiry by investigators into the whereabouts of the 2 men afterwards because they had not been seen after that, and nobody seems to know what happened to them, or if they do know, they are not saying.

§

The Autonomous Economy I - The Strong Man

As a repeat of the same event starts to commence a month later, the rush begins and the melee unfolds. A flood of people start to pour across the river, while a larger brigade of Texans and other assorted Americans, mostly male, descend on the crowd as they arrive. They work very methodically. The groups at the front successfully sink or overturn about 2 out of 3 of the water craft as they approach. The people holding the lines on the other side attempt to pull the sunken or overturned craft back, to little avail. Of the ones that get through, most of these people attempt to run up the embankment, but they are caught by an overwhelming number of catchers that tackle them, or apprehend them by pointing a gun and shouting, 'deten, gota sobre el tierra', meaning 'stop, drop on the ground'; or in some cases they simply punch or tackle them, knocking them to the ground, then tie their hands and feet using plastic cable ties. They are left where they fall after being tied, but afterward, a bus comes and picks them up, where they are trucked to a processing facility. Their medical condition is assessed to be sure they are not injured, then they are given water, or hot beverage, and food. They are interviewed, and vital records made, then their picture is taken along with fingerprints. They are then boarded back on a bus and trucked to the nearest crossing area, where they are unloaded and sent back across the border. In some cases where the crossers are egregious repeat offenders, criminals, or they are deemed as dangerous people, they are taken and processed through the local and state criminal justice system, with various charges designed to imprison them for reasonable duration.

On this particular occasion, one of the rafts ferrying people across was sunk about half way across, and the occupants dumped, requiring them to swim the distance past the many obstacles while Texas cowboys are awaiting their arrival. One of these men washed ashore barely able to move. He lay on the river bank where he was dumped for about 10 minutes muddied and wet. The man attempted to run up the embankment, but was apprehended by the posse, and dropped, then hogtied. He was obviously a white man by appearance and language, with a mustache and wearing a dingy blue business suit, appearing as if he had lived in it for the last several weeks. After he caught his breath and gained some composure, he asked to be taken to see the man heading the operation, so he was obliged.

The man was taken to see Deputy Sheriff Thomas Garza, who recognized who the man is while he stood in front of him.   He is restrained by the cable ties placed on him by the catchers, and quite soaking wet, disheveled, unshaven, and covered in mud with burrs stuck to his hair, his pant legs, and the back of his jacket.   Deputy Garza, with a smirk on his face, as if he is about to bust out in howls of laughter, greets him, "Well I'll smack a monkeys balls, look what washed up on the shore.   People been missing you for about a month mister Blau-stine, where you been?"

Around the room, several of the other officers are laughing and smirking, and making jokes at seeing the man.   Garza grabs a flyer with a picture of the 2 missing men off of his desk, then continues, "Did you get sent to see how the other half lives, down there in cha-cha land, I bet you did, hope your stay was pleasant?   That has happened more often than I can remember you know, quite often in fact, with your friends anyway.   I hope you are not mad!   I, nor the Sheriff can control what everyone does in our county.   I reckon you ended up in the wrong place with your nose in the wrong people's biness; shouldn't do that.   By the way, was the other fellow with you, have you seen him?"

Blaustein angrily retorts, "He was with me in the boat on the way over.   I have not seen him since."

Then in an agitated and angry voice he continues, "I have been a civil rights attorney with the government for many years, and I have seen incredible violations of virtually every law on the books, by every cretan redneck and clan member that holds office anywhere, and I have been assaulted, spit upon, wrongly imprisoned, and called every conceivable name imaginable, but this is the first time that my own civil rights have been so egregiously violated.   I hold you responsible. You allowed me to be exiled from my own country.   This tops it, you deputy, and the sheriff had better prepare for the ass kicking of your life.   Get ready to kiss your badge and your freedom good bye, because I can guarantee that you and the sheriff of this rat-hole of a county are going to regret crossing my organization.   When I get back, you are going to hear from the attorney general himself!"

Garza, sitting on the corner of his desk, while the man continues his rant, picks up his biness card, and tucks it into Blaustein's jacket pocket, then responds by making a mocking jester with his hands, parroting him as he mouths "blah blah blah", then continues, "Get back, you gonna get back to where? You are an illegal border crosser mister Blau-stine, the only place you're going to get back to is the country you illegally entered from, but if by chance you do get back this way again, here is my card, so you know who I am, or if you just wanna chat!"

Garza then gestures to an officer, who comes and escorts Mr. Blaustein, and puts him in a holding tank with a recently nabbed crowd of illegal aliens, who have already been processed, and are waiting to be sent back. As he is being led away he yells, "I am an american citizen, and I am entitled to a phone call!", to which Garza replies, "Well mister Blau-stine, that would be right, but right now, you fall into the category of illegal border crosser, no phone call!"

The American public was in an uproar like never before, and many became radicalized, taking to the streets violently demanding something be done to turn the situation back from the headlong path. Others armed themselves and started taking on the foreign criminal hoards setting up operations. It was like a repeat of Gangland Chicago in the 1930s. A whole new crowd of politician was swept in, so the system was modified, while the old crowd who would not repent from their evil ways and views, were swept out. New hope abounded, that a renewal of the civilization could begin, so a new course was anticipated.

CHAPTER THREE
Simple State Of Siege

[MD Narrative]: Fiscal collapse looming and precipitating rapid economic destruction, with several states and localities being invaded by hostile foreign entities and a weak and ineffective government failing at all levels, the conditions are rife for revolution or outright rebellion, with a movement toward secession and the sovereign independence of large segments of the country. All of these factors are contributing to a low-level state of anarchy in which local and state authorities all but ignore the federal government. It had reached a very critical situation, and the nation may have blown-apart; it certainly was headed that way. Had that commenced, the ultimate destination of our history to this point would not have been altered, however, it would have followed a much different route than what it did. I have witnessed many such close situations over many years, where the decision to go one or the other direction hinged on a few individuals with maybe a single shaky idea or proposal, saying a few persuasive words at a critical time, which averted what may have resulted in a total economic collapse, with the full breakdown of authority and a complete onset of anarchy and domestic civil war.

§

The government as weak as it has been, has let criminals and foreign interests, threaten the security and sovereignty of individual states, cities, and counties. The standard social order rules of the day for dealing with such problems were created under prior, more or less, normal circumstances, but these rules are wholly inadequate to deal with a rapidly worsening situation, where, under the old rules, the problems are intractable and threatening large segments of the nation. The people living under these circumstances, while adhering to the old rules cannot extricate themselves. However, these are not normal circumstances, because of the weakened state of government giving way to a general state of siege, besetting large segments of the country, which grows and threatens the nation as a whole. Courageous men take it upon themselves and ignore the old rules and the politically expedient expected orthodox path, and instead, adopt new rules and the measures necessary for remedy, thereby permanently fixing the new rules and the ultimate course of the nation.

In south Texas, Sheriff Clifford (Buck) Quantrell, who is a 45 year old 6 foot tall white Texan enters the office at about 12 thirty PM carrying his rifle, which he lays on his desk, he is dressed in hunting garb including camouflaged cowboy hat, jacket, and pants. He takes off his outer vest hanging it over his desk chair, then walks over and pours a cup of coffee. He greets the dispatcher who is very busy on the radio coordinating dispatch with the numerous officers busy handling the mop-up of the recent illegal run, "Sheila, how you doing? Deputy Garza about?", he then ganders out the windows into the holding pen observing a full slate of officers busy processing the men caught in the border run.

The dispatcher Sheila, replies "Welcome back sheriff Buck, the last I heard from deputy Garza, he was still out with Walt flying over the river surveying the run."

Buck replies, "Yea I heard this one was bigger than the last, seems he's got it under control. I just drove in and need to go and cleanup, I'll be back in about 2 hours. If you see him, let deputy Garza know I'm back and he needs to come see me will ya?"

The sheriff picks up a few things, then picks up his rifle and heads back out the door, and into his truck, then drives off down the road. He returns later dressed in his sheriffs uniform, and goes in to address the troops. Deputy Garza gets back and greets the sheriff, "Hey Buck, you back already, I didn't expect you till later on. We're just finishing the mop-up. Man you were right about accumulation, I was telling Sheila I estimated what, about ten thousand this time, where we thought about 8 at the end of last month. How'd the huntin go, you bag a buck, Buck?"

Buck replies, "Well deer season doesn't start for 2 weeks numb-nuts, you are supposed to know that aren't you? I would have liked to postpone and go then, but I had some other biness to attend that just happened to be this week. Come on in here, I need to fill you in on it."

Deputy Garza follows the sheriff into his office and sits on a chair in the corner as the sheriff moves around and sits at his desk. "Tell me, how did the burrito round up go, any problems?", Garza replies, "well, none so far, oh except, you are never gonna guess who washed up on the river bank this side?"

Buck then guesses, "Who, DB cooper, or amelia airhead?, hell I don't know deputy, just tell me, I ain't in the mood!", he replies, "That lawyer from the civil rights group who's been missing for the past month, he finally figured out how to hitch a ride back in a taco-boat. Speaking of tacos, you had lunch yet?"

Buck replies, "I assume you know that because you picked him up, or did he just walk in and tell you he was gonna have our asses? No I ain't had lunch yet, lets walk over to Rita's and we can talk."

Garza gets up and walks out of the office and into the hallway with the sheriff behind while he replies, "Ok! Well, he did mention that, but no, he washed up, mud, burrs, sunburned, and smelling like a mexican whore house." On the way out, the sheriff says to Sheila the dispatcher and office manager, "Hey Sheila, the deputy and I are going over to Rita's for lunch. Be back in about 45 minutes!"

On the walk over Buck asks, "Wasn't there 2 of them missing? What about the other one, he with him?", Garza replies, "Nope, no sign of the other one. I asked the one that made it back, mister Blaustine about him. He said that the other man was with him in the boat on the way over, but had not seen him since."

Buck asks, "What did you do with Blau-stine?" Garza replies, "I processed him and sent him back, what'd you expect?" to which buck comments, "Well that sounds pretty clean, I just hope this never comes back on us. Them jew lawyers from up north can be real vindictive SOBs."

Arriving at Rita's Tacorium, they enter and sit down and order their usual fare. The sheriff then fills Garza in revealing that he not only went hunting but also was invited, and so attended a meeting of sheriffs and other law enforcement from 5 states about the growing problem of heavily armed drug cartels taking up root in communities, building drug manufacturing plants, and distribution infrastructure, recruiting hardened inner-city gang members to act as enforcers in virgin communities, and young unemployed local men to work the plants.

Sheriff Buck comments, "The problem is serious and getting worse across a much larger region than we know. Small towns and cities all around the country are seeing it. They are being ripped apart, like we were 20 years ago before we got smart, with the homicides, the criminals, burglaries, assaults and the community being split between factions for and against it. They mentioned 3 local officers caught and prosecuted for taking bribes in a small out of the way little berg that no one ever heard of. Just the beginning of the infiltration, and you know it don't take long after the first few take up."

He continues, "There were present, some from the statehouses of 2 states including Texas, warning of a coming failed state like across the rio, and threatening secession. This is the statehouse saying that...; back into an autonomous region and a new union with others entirely separated from the current; if something is not done, and soon. They are not getting any support from the feds at all, for drug interdiction, or the flood of illegals, and they are suffering continuous and unrelenting statehouse protests right on their door steps. They complained the feds are holding back law enforcement from acting to stem the tide, of course we have seen that for a long time. Welcome to the club, they have been holding the door open on us for the last 40 years, was my comment to them. But, this is finally serious talk, there may be a few states leave soon, which could start a stampede of the rest."

Continuing, "You know the DFIA; Drug and Firearms Interdiction Agency? We met and were briefed by several active members and members of a few other federal agencies who in fact, said they were in direct insubordination for being there. I'm sure they were flexing to sound tough, but they referred to themselves as a rogue or black faction. They said their country is under siege, and needing their help, but their agencies are running nearly on empty, and there is absolutely no leadership from the executive, but only obstruction. One of the men there explained that they were operating under some black or covert protocol, developed and authorized in the 1950s under a possible scenario such as this, that even the executive is unaware of."

Buck continuing, "If you ask me, it is no small thing, nor is it a surprise that these agencies are in revolt and acting insubordinate. It reveals there is a low-level state of rebellion with segments of the government-in-power in rebellion against itself. The danger is that it devour itself by acting contrary to its mandate, and the interest of the nation it is supposed to serve and protect. I am not at all surprised at this development. The only thing that surprises me, is that it took this long to get here. Almost seems we have been pushed to this point. I would never have believed, and I have never seen a nation so bent on its own self destruction like this!"

Continuing, "In any case, they have asked for our help."

The cartels have developed direct air routes in and out of their compounds, some located in residential neighborhoods in white middle class communities. They are buying entire blocks of buildings in the warehouse districts of small midwestern towns, including old factories, who employ thousands of locals, just as back in the days when these small town factories were humming, busy cranking out manufactured consumer and industrial products, and even mothballed elementary or high schools. Now those factories are cranking out narcotics, and the employees have been forced to act contrary to their own safety and security, and that of their communities. The ones discussed, criminal enterprises, well organized and funded in every way and have been known to protect their operations with heavy arms including artillery, mortars, large caliber automatic weapons, and armored vehicles. They have decided to stare down the American government, in defiance, and a direct challenge to the authority of the states they are operating in. In some communities, they have taken control of small airports to fly the drug materials in and out of the areas.

Buck continues, "I surmised that what they mean to do, and in fact they said, several governors have asked for intervention. Supposedly the governors of 5 or 6 states met and decided they wanted to nip this in the bud, but not open a can of worms, by direct military involvement, or by handling this with national guard, only local and interstate police and whatever federal help is forthcoming. There is a general multi-state task force plan in effect which is supposed to be a secret, but the plan is to hit and shut down, or completely wipe-out the 6 largest of these compounds on sunday when they believe there will not be any workers there, and you deputy Garza, are invited."

The Autonomous Economy I - The Strong Man

The effort, while it is still a civilian operation, has recruited the cooperation of a small contingent of special-forces army units, with the involvement of a few generals, including a young 3 star General Jerome Cranket, being one of them willing to stand by but lend support and backstop the operation if necessary. He reveals further, "I wanna tell you something else too, they have said this, but it is only in the planning. There is a fallback contingent operation in which, if the economic, security, and lack of control conditions, which are directly responsible for allowing lawless enterprises to take root, do not improve soon, there is consensus among many that will involve a direct military invasion and control of a large portion of mexico, and at the very least, they will occupy a strip that extends a hundred miles or more deep into mexican territory across the entire length of the border, and this is being planned and will be carried out by an interstate directive, outside of the authority of the federal government. That means states in direct rebellion against the government."

Continuing, "In any case, the rules have changed and when it comes to those suspected of being in the country illegally, conducting criminal activities, dealing drugs, threatening the domestic peace and safety, the law, or what have you, the usual civil procedures no longer apply. No new policies have authorized any changes, but nevertheless, they have changed and for the most part, law enforcement may act much more forcefully, in whatever way is necessary to restore order and remove the threat. There is no appeal, no asylum, no advisement of rights. If you are a bad dude and you are caught, you do not pass go, you do not collect 2 hundred dollars, you go straight to hell."

He continues, "An invasion of Mexico would certainly make our lives easier, and some of those present believed the local mexican authorities in the border area would welcome and help facilitate the intervention by American forces, even if that means the loss of Mexican territory to America. He said, in the northern part of mexico, most people are oppressed by cartels, and would rather be American than Mexican, and mistakenly believe that an American takeover would make them so."

Both Sheriff Buck Quantrell, and his Deputy Tom Garza, with a contingency of their best men head off to participate in the organized assault against intrenched cartel compounds in the states of Texas, and Oklahoma.   Arriving at the county sheriffs office in a small northern town near the border between Texas and Oklahoma, they walk in and are greeted by another group who bring them in and introduce them to others attending for the same reasons. Then they are briefed on the operation, by the sheriff of the county where the operation is taking place in their role supporting the effort, and told of a more thorough forthcoming briefing by the DFIA.

The men are checking their weapons while standing around talking when a representative from the DFIA walks in and addresses the group, "Gentlemen, thank you for answering the call and coming. My name is Brian McDonald, I am a field agent with DFIA and sitting behind me is agent Brad Caruth, also DFIA, but we are not acting officially in that capacity. For our purposes today, and as far as you are concerned, we are acting as civilian consultants. We have been asked to consult on this operation by a contingent of governors, of 5 or is it 6 states?; the number slips me for the moment. Anyway, that's why we are here today. You are here because you have the skill and expertise needed for this operation. In particular, because this action falls outside of being sanctioned by the agency in which agent Caruth and I both work. We are officially acting as civilians, and may be in direct insubordination, however, that is not a concern we are taking into account at this time. As a matter of fact, our agency is restricted by those above the director, from participating, but we are here taking on the assignment because, we have been asked by several states that need our help, and we are answering the call just as you are. Frankly, the federal government has refused to act, and even acts to restrain its own agencies. While the reasons for that are not clear, it is another battle that we are not here to fight today. Any questions so far?"

He stops and waits for questions from the 50 or so people present in the room, then continues, "The larger operation is going to take down 6 targets in the midsts of our heartland this weekend, and we in this room are going to hit one of the 6 compounds that sadly we could never before now have imagined we'd have to deal with; but it is here as a result of policy expressly intended and designed for failure, and it is our job to eliminate the threat, in direct defiance of policy intent. There are 5 other operations that will also commence today simultaneously. Our target is a warehouse in an industrial part of the city, about 2 miles away from any heavily trafficked areas. You will be given locale, as we approach."

Continuing, "Local law enforcement is already on hand shutting down traffic in and out of the area. We estimate there may be 25 to 30 locals present on a shift, so we need to be careful to avoid any civilian collateral on this. We have conducted tactical surveillance and gained intelligence that shows they have an armored vehicle parked inside the compound and we believe it is intended to transport cartel honchos out in case of any significant heat. It will head to the adjacent air-strip. We know they have RPGs so we need to take precautions. Preliminary intelligence suggest they may, and I stress maybe, because we have not confirmed, whether they have obtained mortar rounds, and at least 1 m2 50 cal. Now I realize many of you may not have ever dealt with anything of this nature, so you may be asking why the local sheriff, sheriffs from other states, and the state police are being asked to provide support in this operation, which by every textbook measure looks like it should be a military or special forces operation, and the reason is that it is on American soil. So I would like to ask if any of you civilian officers have any military experience?, ...you, where did you serve and for how long, marines, 6 years? and you, marines also, 10 years? I see a few others, it looks like maybe half of you have had some exposure to military, so you have an idea what you're up against. If you would, for the purpose of planning and coordination, those of you that have any military experience, please give Agent Caruth your name, contact information, any special training and what branch, rank, and how long you served. Thank you gentlemen, we move in 1 hour, at eleven hundred twenty."

§

The cartel's complex is busy with 2 mexican men sitting in front of a group of 15 or so job applicants from the local area awaiting their turn for an interview outside on bleachers. Both of the men are in their thirties. One of the men is dressed in a business suit, and the other in genes and a fancy western shirt with a bolo tie, cowboy hat, and cowboy boots. He is holding a microphone and speaking to the group in broken english, "Fabricación Nacional del Norte is welcome joo for working to help combine peoples and business in mehico with takesas. We want to help people in takesas work and earn money and welcome mehico to jour country. If joo are hired, joo will be taken to tour and learn jour work job. Por favor, go to sign forms and wait here for interview."

Both of the men are high ranking members of the Chimali Cartel, based in north-central Mexico, who regularly practice a form of cultic blood rituals and black magic involving human sacrifice, which they believe will protect them from their rivals and the authorities anywhere they operate by covering them with a shield. Chimali is the Aztec word for shield. Chimalis' are a particularly ruthless group of killers, who's past history indicates willingness to engage in multiple inhuman atrocities of torture and indiscriminate mass murder and mayhem, even in their own backyard amongst the people they grew up with, and family, and most certainly amongst the people of a foreign land. Now they have expanded into the country and are in a race with a rival cartel to gain as much territory as possible.

The prospects for these jobs was a welcome sign to the local community which has been completely ravaged by the economy for the last decade, with no signs of recovering soon. The group is interviewed one by one as best can be done considering the language barrier. The jobs did not require any previous qualifications which signaled a red flag when one of the local county labor and regulation compliance offices learned of the openings. In their application for license and waivers, the company listed manufacturing and fabrication of parts as the business conducted, however, it is unusual for a manufacturing facility to advertise so many new job openings without any skill requirements listed. An officer went down to the location before any hiring commenced, to inquire about the requirement and was assured by the plant manager that this was indeed the case, no skill necessary; employees were to learn on the job. The officer then inquired about what it was they manufactured, and was told they manufacture industrial supplies, and was shown a few samples of rubber o-rings, and small clips, and machined metal parts. He asked to see the work floor and machinery, and was shown a large empty room with numerous tables and workstations setup, a few conveyer systems to move materials around, large mixing vats, a complete chemical lab including exotic analysis equipment, a room full of tanks of gases and drums of chemicals, some storage bins and cabinets, plenty of packaging, labeling, and shipping materials, weighing and sorting equipment, but no fabrication machinery, nor manufacturing hand tools. The officer filed a report suspecting the use was for large scale drug manufacturing, so several months later, the local sheriff finally sent an undercover officer in to interview and be hired, confirming the operation was cartel related, and prompting the raid.

The mexicans had reworked the plant spending a considerable amount, then because the facility was on the outskirts of the little town, and more or less surrounded by wooded areas, they were able to covertly cut a path through the woods to a large adjacent empty cleared area that at one time was farm land. They installed a small airstrip, which they believed the county would not find out about.

§

Congressmen Cy Schuman and Henry Stumble along with several other members of congress are attending an event in the nations capital involving education and the welfare of youth and the community. They both speak of the large budget programs which they personally support, have sponsored or co-sponsored in congress, or are involved with, and which they personally believe-in; and being politicians they have mastered the crowd-language most effectively employed in inspiring the simpleton masses using worthless gibberish, double-talk, and platitudes about their programs with such drivel as, the program will, '...build strong and healthy environments for the nurture and upbringing of children, and strengthen all of us and our communities through unity', '...strengthening our world through cooperation', and, '...ensuring our unity through the strength of healthy environments', blah blah blah! Volumes of this tripe have less substance than a single popcorn fart, and are designed to solicit the adoration and support of the bubble-headed vapid moron masses, possessing the depth of humanity in themselves consistent with that of the common garden toad. At this juncture, it was unfortunate that the nation was populated by a majority with this mentality.

Both congressmen are ardent advocates of spending large amounts of public money on worthless local and national programs that get the community involved in the activities of youth. They both ostensibly advocate positions as staunch foes of drugs and the influence of the drug culture, citing the toll it has taken on communities and the nation. They speak along with several others, and in the course of their message, they introduce the crowd to their special friends including one Eloy Duarte Murguia, and Pablo Sanchez, who are ostensibly, the president and manager of a Mexican multinational corporation that has expanded its business from Mexico into America in various locations. They operate as Fabricación Nacional del Norte in Texas, and by other names elsewhere. The event is held indoor with a large number of people in attendance, and both Congressman are sitting in the VIP section next to the stage, were they are scheduled to speak, and to introduce some special guests, with whom they intend to lavish accolades upon, to demonstrate how much good they are doing.

While Congressmen Stumble is waiting his turn to be invited up to speak, he is in a hall talking with both of the men, "I don't care what your compadre from Kansas told you, and I don't care if you are having trouble with a local licensing authority or a county whatever, I don't care what your problems are, they are not my problems, and I especially don't deal with problems that are not even in my state. You boys have to clean up your own damn messes, and not the way you do it in Mexico, this is not Mexico, so don't even go there. You can not think you can drag me in whenever you have problems. You and your enterprise are a contingent risk that I am not sure balances out in the calculation, and if that risk reaches an unacceptable level, I will be forced to re-evaluate, and eliminate the potential liability. I am not your damn baby sitter, nor am I your tour guide here in the states, am I making myself clear? Look, if you want my help, you had better start listening to what I am telling you! Now I'm going to go give a speech; which one of you is going to present? Decide, then get ready for when I call you up, say a few words, then get the hell out of here. Now, you said you have something for me?"

The man starts to pull out an envelope, and the congressman stops, grabbing his arm before he produces the package, "Are you crazy or just stupid? You cant do that here, in front of people. I told you this isn't mexico. After you are done, go see my assistant, Perry. Give it to him. Make sure its what we agreed on."

The Congressmen is introduced by the host and gets up to give his long winded speech, which he barely starts before he immediately introduces his guests who are on a very tight schedule, as they have been told to get the hell out of dodge, "We have seen this nation torn apart by drugs, and it has taken its toll on families and children all across this great nation. We believe that we can have an effect in countering this terrible scourge, but we can not do it alone. It will take your help as well, and we will have to elicit the help from many quarters, and many, many others in the struggle including some who have a story to tell like these 2 dear gentleman right here!", the congressmen points to them, then gestures to the men to step up to the podium and speak.

One of the men, Pablo Sanchez, gets up and walks to the podium to speak while Congressmen Stumble continues, "Come on folks, lets give a big hand to this gentleman, mister Pablo Sanchez, come on up here Pablo." The Congressmen then states, "It has only been a short while that I have gotten to know this man, but he has a story of struggle, and adversity that would bring a tear to your eye, and swell your heart if you heard it. We in America know what that is all about, but we could all take a lesson from this man, I'll tell you that! But, you know, in-spite of the hardships he has endured, he has raised himself up, to overcome tremendous adversity, to become successful, and has shown us that it is still possible, and because of that, he has told me personally that he would like to make a very generous donation on behalf of his company, to local organizations such as this one, to help those ravaged in the war against the scourge of drugs in our communities. Mister Sanchez, the microphone is yours!"

Sanchez starts, "Gracias, mister congressman Stoomble. I am Pablo Sanchez, Gerente General of Fabricación Nacional del Norte in Takesas. Well, I only like to say that even though I did not grow up in jour contry, I have liked being here, and making business from Mehico in Takesas. It has been working fine, and we have feefty eemploys, that has jobs, and making money. So, I don't like when people say to make drugs legal, because drugs are bad. I come from Mehico where drugs dealers kills many people almost every day. It makes the life on people very hard, and they are mostly afraid every day, when drugs dealers are in the noos. So I am giving thanks back to America with sponsoring with a check for feefteen thousands dollars for this drugs fighting organizacion, gracias and thanks to you!"

§

About 8 months into the operation, when they had hired around 80 local people, there were rumors spread around the area about what was being manufactured at the plant. On the one hand, people were very glad to see some paychecks and finally a little money started to flow in and around the town. On the other hand, people were not stupid, because they knew that it would not last, but it was bound to end soon and very badly at that. The majority of the population did not really want the scourge in their midsts. Even though the economy had soured, they believed that it would eventually get better, so did not want a bad legacy to taint the name of the small conservative religious community, nor support the poisoning of what is left of the once proud but waining nation. The town was at odds with itself, with some arguing that as long as it was able to operate and make money for the area, it was doing good. It was like the days of liquor prohibition, when proud southern people made the liquor in stills, and were darned good at it, while the government that tried to stop it was evil and overbearing.

They argued that prohibition and the demand, for the production of white lightning, bourbon, and whiskey was actually good for those communities. Those on the other side argued that it was nothing like making liquor. Whiskey may have had some problems, but it did not do anything to a community like what addiction to narcotics such as crack cocaine, or meth-amphetamine did, and the criminal activities that were associated were much worse. Property crime, murder, intimidation, corruption. In Mexico, since the 1980s, there were hundreds of thousands of murders, and atrocities where as many as 100 people from a small town were tortured and murdered, having their heads cut off, and dumped out in the open, just to make a point by these ruthless cartels, and now it was here. Then there was looming, the corruption of officials and respected members of the society, and the statehouse at large. Mexico had never really amounted to much, but when the drug cartels gained power, they took over the reigns of a very weak government, to completely control the country. The majority of the people agreed, and they held city counsel meetings, and motions were adopted that demanded the scourge be removed at once, and their cries had reached the governor. That is when the intimidation started, with numerous murders of local authorities and prominent individuals and their family members.

Several of the young men that had been hired at the plant were taken on a tour and started in the shipping and receiving area, and their activities and knowledge were strictly controlled. They were gradually psychologically conditioned and groomed by promises of wealth and power in the structure of the organization. They gradually assumed more knowledge and responsibility in the illicit activities, but had to keep their mouths shut, which was something Americans had never learned to do very well. Several were allowed to oversee and help plan distribution routes, recruitment, set up of other facilities, and enforcement. Once the insidious tentacles of the organization took hold, it grew like a cancer at an alarming rate, by starting other geographically disperse facilities, and developing a network of compromised officials paid to look the other way. Other recruits were trained and groomed by several of these. They processed raw materials into crack cocaine, powder cocaine, and meth, on an industrial scale, monitored quality control, packaged it in precisely measured units, and even applied brand labels, then shipped it right out to distribution warehouses and on through the distribution network to the various street level dealers throughout the country.

With some of the profit from the drug trade, the cartels became covertly involved with propaganda, and through financial support, played activist on both sides of the issues concerning the possession and use of narcotics and drugs in society at-large. The aim was to influence society into favoring laws that favored the business of the cartels, and to a much lesser extent, they believed it would promote the image of themselves as benevolent heroes of the people, who would protect them when necessary; a tactic that had some success in South America, but little effect here. They mainly supported voices demanding the scourge be removed, while making sure to counter the voices on the opposite side of the issue that argued these substances should be made legal, fearing what that would do to their enterprise. This support was undertaken in many ways, but of particular interest, was to support those with political ambitions that called for stiff penalties and sanctions against the sale, distribution, and use of drugs, which would on the surface seem counter-intuitive, but the simple facts were that many ignorant voices calling for the legalization of illicit drugs had been gaining ground and growing louder for decades, which was a terrible threat to the cartels.

The new approach and ground gained by moving production across the border, was a step forward, and eliminated the overhead costs associated with smuggling the finished product into the country. The risks were many but one of the most damaging threats was the risk of being squeezed completely out of the business they introduced into the country by the Americans they were dependent on for the work and protection. After-all, America, despite the teetering economy, was still orders of magnitude larger than any other market they could hope to penetrate so thoroughly, and America had already mastered the illicit underworld game of cops and criminals in the 1930s, with bootleggers and speakeasies, so they had been there first and were already adept at the game. The average American criminal could easily and considerably exceed the prowess of the average Mexican, in virtually every measure when it came to innovation and pioneering new and creative ways, with the possible exception of the depths of ruthless depravity and violence; of that, they were equally capable.

If Mexican cartels could successfully operate on American soil, given time and enough money, Americans themselves would certainly completely eliminate them, then streamline the illicit drug operations to precision levels of efficiency and productivity never seen before, anywhere at anytime. Now they only needed to smuggle in the raw materials or legally obtain it. At first it was easy to simply fly a load of coca leaves, or chemicals up from mexico, but later they shipped in through the ports along the gulf coast and simply trucked it in. Business had begun to really take off, and looking like it might fulfill the Mexican to American cartel transition envisioned in short order, when the raids happened and shut the entire operations of 2 rival cartels down cold!

§

The raid commenced from an obscure staging area about 7 miles outside of the city so as not to tip off the targets of the operation. They arrived in the area promptly at the designated time, and the officers swarmed into the community early in the day shutting down roads in and out of the area. The officers quickly descended on the compound and rushed through the doors, and arrested all persons present. The purported .50 cal m2 machine gun had been setup on the roof and was used to fire on the police in order to pin them down, while allowing the honchos to enter into an armored personnel carrier vehicle housed inside the loading area of the plant. The m2 was finally subdued and the shooter killed by SWAT, who got into the building quickly then pinpointed the location of the machine gun on the roof through the sound it made, then firing from underneath through the ceiling toward the location, but not before allowing 6 high level cartel members to pile into the unit and drive away out the back, crashing through the wall.

They first attempted to escape through a reinforced steel loading door, which would not open because the security system was engaged, and because they were in a hurry, so they fumbled around attempting to open it, but were unsuccessful. They subsequently attempted to crash out by driving the APC through the steel door which was built to their own specifications, making it impenetrable by such means, but the vehicle ran into the door at considerable speed denting it and mangling the door's drive mechanism, making it impossible to open, thus trapping themselves inside. The vehicle simply bounced off of the door, doing considerable damage to its front, and the door. The driver then backed up and attempted to try again, but this time he drove head-on into the block section of the wall, weakening it, then on the 3rd attempt, the vehicle broke through. By this time police had entered the loading area and began to shoot at the vehicle as it broke out.

It started down the passage way cut into the woods that led to the airstrip. The passage went straight for about a quarter mile, then made a very sharp left. This was to obscure that the passage actually went anywhere when observing visually from ground level. However, police had done an arial surveillance ahead of time, so they knew where it went and were waiting there in case the cartel members made it to the airstrip. Upon arriving into the area of the airstrip, they proceeded to the south end of the field where they had stored a small plane and 2 helicopters. The aircraft were already in possession of the police, so there was no way for the desperadoes to escape, but they charged ahead toward the police at full speed, and the signal was given to the army unit to fire upon the armored vehicle. The army supporting unit fired an armor piercing sabot round into the vehicle which disabled and stopped it. The round also killed all the occupants.

The task force then mopped-up the operation by searching the compound to discover a plethora of powerful and exotic military weapons including 200 mortar rounds and tubes, numerous fully automatic military small-arms like assault rifles and hand guns; 3 20-

millimeter auto-canons intended for use in bringing down aircraft; and several shoulder fired missiles. The materials and documents recovered from the raid, along with that from the other raids, rendered intelligence of the extent of both cartel's entire operations inside the country. The intelligence gained, allowed the authorities under the DFIA multi-state joint-task-force to disable the smaller fledgeling operations of both cartels involved, and to eventually apprehend all people associated with the operations still in the country, setting the cartels back many years, dissuading them from attempting to operate on American soil at such a high level ever again.

The task-force issued a report back to the regional organization that commissioned the action, it was given to the original group commissioned by the governors of 6 states. In the report, they named the compound in north Texas as being operated by the Chimali Cartel, based in north-central Mexico, and named all the criminal participants, and what became of them, both Mexican and American. The report then made news headlines, and was confirmed by local reporting, and so in various places around the country, it spurred on support and adoration for the heros of the operation, applauding the courage, audacity, and defiance of the group including the governors involved. It put further pressure on a very weakened central government teetering on the brink of certain disaster. It gave considerable hope to an America very much in-doubt about its future. It placed the official inept, incompetent, and increasingly worthless government in the nation's capital, in sharp contrast with a group that saw what needed to be done, and did not worry about all the bureaucracy, but acted where it was needed, and actually accomplished something, like in a previous, more functioning era. It engendered hope, that if push comes to shove, there are still some leaders that have the nerve and fortitude to carry out the express will of the people, to go the distance and do what circumstances demand, regardless of the bureaucracy standing in the way. Many heroes were created and became overnight celebrities, including Sheriff Clifford Quantrell, and Deputy Sheriff Thomas Garza. The stories of Agents Brian McDonald, and Brad Caruth were also told and they were asked to comment but both declined to talk about the events even though they both had a hand in organizing and planning the operation, and authoring the report.

The Autonomous Economy I - The Strong Man

Finally it stood as a warning to the government in power, and a message to the people, that maybe it had become too large and inefficient, too old, inflexible, and established in its ways, too tired and inept to function, and way too corrupt. While the action of this group began to show the promise of something new, something exciting, something hopeful; that now many began to see, were better ways, with the glimpse of a fresh morning. If the nation as a whole did not get its act together, in accordance with the vision and demands of the people affected in just 6 states, there was an escape for them, along with all those in other states, suffering the same insults, who are willing to forgo any further cooperation with the existing deteriorating union, and join a new one. To establish a new nation from the old with the likelihood of many more existing states to follow and join the new. A civil divorce.

Likewise, it also revealed that as in America's past, both it and Mexico, like 2 children, they once played together in a very wild west, and whereas America had grown up to become considerably more civilized and successful for a time, Mexico had not, and now America was beginning to revert back to its old form in that respect, but now, even that may be preferable to the oppressive overbearing state of control by an inept and failing central government. In any case, even if only for a brief moment, it meant that at least for the interim, the people were once again in charge of the affairs of the nation.

CHAPTER FOUR
Automation Marshall Plan

[MD Narrative]: As with all things, eventually the fog started to lift and the light started to shine. It was a known axiom that massive increases in productivity, or the corresponding drop in the cost of production of goods and services, could increase the general prosperity, by offsetting the inflationary effects of continuous monetary stimulation on a debt driven economy, thus curbing some of the problems associated with financial decline. The reasoning put forward was that if a little worked, a lot should solve the crisis.

§

The movement toward massively increasing the number of 3rd world immigrants was less a solution to the nations financial woes, and more a political power-grab that was doomed to total failure by the defiance of a nation weary of the games and abuses by government and special interest groups that seek political or financial advantage at the expense of the nation at large. However, the phenomena was not altogether without positive effect. Indeed, it started a conversation and thereby focused a spotlight on vital debate. There was congressional testimony from economic experts involved in the creation of capital and dynamic systems geared for economic growth, which clearly voiced the economic advantage gained from high productivity.

Conditions have continued to worsen through the later part of the year and into the early part of the following year, but still there is no reprieve in sight. People across the nation have been preparing for disaster for the better part of 9 years, building shelters, storing up food, acquiring weapons for personal defense, erecting neighborhood armored and manned fortresses, fearing the threat of a desperate and hungry mob, believing the end of the world, or at least their world, is imminent. Nearing the precipice of complete fiscal meltdown, which means the shuttering of virtually all government offices and services, creating a power vacuum which drives para-military fringe groups, and the accelerating collapse of the nations economy, the people's house is still in session on capital hill, with testimony once again before Congressmen Stumble's committee, with the congressman presiding. It seems the short sighted bureaucratic planned invasion from the 3rd world has been a colossal failure since the people of several states have taken the matter into their own hands and turned back the invading mob stampeding across the border, and have now gone beyond threatening secession, but have defied the fiat of US 'central planning.' Having gotten a taste of defiance, and knowing the weakness of the federal government, they are in a mood for revolution. Many disparate groups formed out of desperation, without particular direction, are in the process of drafting ad-hoc committees, and planning for a total restructuring, which includes a new and separate union from the current, with a new constitution. In essence, a new order from the old is about to emerge. The first of several delegation sessions are planned to commence in the next 6 months for the restructuring, so delegations are being assembled. Now the current nation faces a possible widespread rebellion and the dissolution of the empire, as states deliberate going their own way, forming a new union, disengaging from the current ship of state which is rapidly going down.

Although on capital hill, the congressional sessions continue, many members of the house and senate do not even bother showing up. A large number of congressmen and senators from several of the exiting states have been recalled by those states, and that list grows weekly. In some ways this makes the process easier due to the critical nature of the mission, and the fact that infighting has ceased. Petty squabbles are not tolerated, and no one engages. It seems that men learn to drop their differences and work together on their common vital interests only after the problems which separate them, have grown to unmanageable crisis proportions. Once they stare into the abyss, and disaster stares back, where they face loosing everything, they are more willing to work with even those considered enemies.

The group testifying is small, composing 3 gentlemen from business and industry. The witness, Greg Tishman, a 45 year old economist associated with a university in Texas and a consultant to industry, is testifying to the committee attempting to explain the advantage, and pitfalls of cheap labor, "Increases in cheap labor resources can add to the productivity of the nation. If the cost to produce things diminishes, more can be created, which is growth by definition."

He is interrupted by a congressman from a midwestern state, "Sir, you have been asked here today to testify on how to solve the nations crisis, which at this point, does not include bringing unwanted immigrants into the mix. Have you not been paying attention to news reports? Have you not seen the rebellion right at our doorstep?"

The witness protests, "Congressmen, sir, please, let me continue, I am not here to advocate for immigration which will not solve our problems anyway. I am really here to hopefully demonstrate how I think the problem may be addressed, and I am accompanied today by a few gentlemen, a team if you will, whom I believe hold the answer. This team represents the solution, and will start a process toward that solution, so please bear with me, let me speak, then you can hear from them."

The congressmen retorts, "Sir, if you have an answer to this crisis, you have everyone's attention, please proceed."

Mr. Tishman continues, "The act of increasing cheap labor by itself, does not increase productivity on a broad base, but instead only contributes to the cumulative decline because it ignores the fact that labor coming from the 3rd world is composed of human beings, and therefore comes with a human cost, which is mostly borne by the already overburdened public, while the advantage and benefit of labor cost reduction, largely only accrue to very narrow interests. For a solution to come from increased productivity, the application must be very broad-based, and that is not possible with a 3rd world workforce less able to produce than the average of the domestic workforce. A 3rd world worker may be cheaper to employ in menial labor, but by and large there is no economic advantage when in the society in which they are employed, they are not able to function at a level high enough to earn their own keep, or such as where the work requirement calls for higher education and skills which they do not possess! Nonetheless, the evidence is clear, and we are convinced that increased productivity is a vital key to moving forward again, but make no mistake, gains in productivity can only come from advances in automation and technology, not human labor, wherein the costs are contained and diminishing!"

Continuing, Mr. Tishman then steps out on a limb and with language some might consider contemptuous, as he reads his statement and expounds on his conviction from a political viewpoint, "Please bear with me as I read a prepared statement. Our problems seem economic, but in reality, they are almost purely political. Indeed bad policy is always at the heart of crisis, and this is no exception. Many Americans, heretofore have chosen to ignore the real basis for previous centuries of economic growth and prosperity, but are now forced to face the truth. Contrary to the lies put forward by ambitious politicians co-opted and directed by elitists, big money and their minions, along with social engineers, and the myriad of self appointed oracles of civilization, who have propagated and maintained such false doctrines as…, *'the degree of prosperity, and economic security in an advanced society, is proportionate to the amount of consumption.'*, and the old failed Keynesian canard that…, *'Government must spend whatever is required to fill the gap between capacity, and utilization below a certain level.'* It has taken 100 years or so, but this one especially has finally proven its falsehood. These have never been sound economic principles, so they also make very bad policy."

He continues, "It has become apparent that the more the government spends, the greater the gap becomes, even to near the entire range of the gap. Just so you understand, I am referring to the gap between economic capacity and utilization, which many of you equate as unemployment or a drop in economic output. So the logical outcome of continuing the Keynesian lie is that government spending will encompass all economic output. By this I mean, there will be no production by private means, no private industry, no individual initiative or enterprise, and no further creation of capital; only slave labor, and stewardship by government apparatchik and bureaucrats, which decades before was known as the failed communist system. Gentlemen, I believe you are all aware that, that is where we are headed, and the reason for where we have arrived today."

He continues, "Congressmen, I believe we may be entering an era in which it has finally begun to dawn on the majority in business and industry, that our collective prosperity has infinitely more to do with what we produce than what we consume. It is undisputedly clear from a purely economic standpoint, that *...domestic production equals domestic wealth!*; and while that has always been a bedrock principle since the beginning of time, in recent decades, the opposite has been erroneously taught, and so believed. It is a flawed doctrine which said that production anywhere in the world makes no difference; or that the benefit of global production would always accrue to Americans, because we consume, and further, likewise, it can be demonstrated to be no more advantageous to the consumer for automated machinery to produce in developing countries than in the developed world, the labor factor notwithstanding. The relative cheap labor advantage, versus the resultant dislocation of workers and the political instability created is the same, whether production is here or abroad. It makes no difference if a widget is produced by automated machinery in China, India, or the United States, it is still the same widget produced at virtually the same cost. The Chinese are good workers and efficient, but the automated machines they use to make a product, and the product itself are exactly the same, whether made in China or not, and the same applies to America."

Mr. Tishman continued to say many things advocating the government adopt policies that eliminate any unnecessary spending, and foster the development of automation and productive technologies for industry, then he finished, and the next witness was called upon. Even though Texas has recently recalled their representatives from the nations capital, nonetheless, some have continued on.

A congressmen from Texas introduces the next witness, Boirix Dalgleish, "Mister Dalgleish, I understand from the testimony of your colleagues that you have a plan, which has been described as something like a Marshall-plan, for rescuing the nation from the mess we find ourselves in? Of course 'Marshall plan' is referring to the plan that rebuilt Europe after the last great war. Now, I have sat and heard the testimony of many of America's finest citizens, who have spoken on what runs the gamut of the problems we face. I have heard stories and schemes that range from just plain naive, or short sighted, to what can only be described as coming from somewhere off of this planet, and amount to literal insanity, for lack of a better way to put it, for the better part of 2 years now, and for all that time, and all those words, I have not heard much of anything that gives me much hope that the problem can even be solved, that is, until this week when I started hearing from your colleague mister Tishman. Something just struck a chord with me, and others on this committee, something rings true and makes me believe ya'all are on the right track. Sir, if you can help us dig our way out of this mess, I am confident the nation will owe you a tremendous debt of gratitude. I would like you to know that I am on my own time here, thats how important this is."

Mr. Dalgleish responds, "Congressman, first of all I'd like to thank you, and the committee for the vote of confidence, but I'm not sure I would characterize our plan as like the Marshall plan. Mister Tishman laid out the philosophy at the heart of what we have to do, and what we think you need to do. It is simple, we can solve the productivity problem, but you need to solve the political problems. We…, no, I, think our entire process of trade and fiscal restraint need to be restructured along the line of what he laid out."

The Congressmen interrupts him, "Yes, I think we have all heard and understand that point, and he nor you are the first to tell us about fiscal restraint, or fair trade policy, we are not altogether ignorant up here, although it may seem that way to some. What we need to know, is how the nation can begin to work its way out of the financial malaise it finds itself in. We have debt that we cannot pay unless we can find a way to grow faster than the debt mounts, and without having to inflate ourselves to oblivion, or sell the entire country off."

Boirix Dalgleish responds, "Congressman, I cannot tell you how to grow at the pace you believe we need to, all I can tell you is that we believe that we can greatly improve the state of economic output. To speed up the current rate of production, to increase productivity significantly, by developing technology to accelerate the process of what is currently done by human labor, by artificially facilitating and enhancing the labor component for most industrial, commercial and domestic services, which means, we can build machines that can produce at a significantly faster rate than their human counterparts, with better accuracy, and reduce the overall costs associated with human labor significantly. Now, this becomes a political issue because, we are talking about the possibility of replacing jobs with technology, and I know that can be a sensitive political issue, but I am not talking about an immediate impact. In fact, it may take as much as 15 years for the technology, and the adoption process to really take root. We have the components of the technology now. What we lack is the time and money to put it all together, but I believe that once that takes place, this technology will bear fruit very quickly in relative terms."

He continues but is interrupted by another congressman, "Mister Dalgleish, may I interrupt for a minute sir. You say that this process may take as much as 15 years before it bears fruit? I respectfully submit that we do not have 15 years to wait, but the nation cries out for an answer now. I do not see us lasting for 15 years for a solution to the situation bearing down upon us now. Are you aware of, or rather, can you explain how this process may be sped up?"

Mr. Dalgleish continues, "Congressman, you ask a question that I can not really answer. In my opinion, the short answer is that we can only do everything we can to speed up the process, but that is all we can do. I'd also ask, what choice do we have? We must do everything in our power to improve the situation, and pray that God will do the rest. It is my belief that much of the problem we face with a restless nation stems from a lack of direction, and leadership, and the confidence that comes from knowing that someone has a handle on the situation. Give them that confidence, and it may give us the time necessary. But, once started, it must be completed; we can not fail to deliver. We must demonstrate progress at every opportunity. It is my belief that if we start down the road, and signal to the nation that there is hope, and we have identified the problem, and have started down a course to correct it, and can demonstrate that fact and answer the skeptics; that action alone may give us the wiggle room necessary."

Dalgleish, Tishman and team impressed many in the house of representatives who voted affirmatively to pass a measure creating a national initiative. The team then went on to testify in the senate, and were invited to assist in drafting legislation to implement the plan which was publicized and described as being 'A Marshal plan to rebuild the US economy', or like JFK's race to the moon for automation and productivity. On the precipice of a disaster, and having the previous 60 years, steadfastly allowed America's productive capacity sent abroad, the US government finally passed **The**

Automation and Productivity Act of 2019, which created massive incentive for investing heavily into automation technology as it applies to manufacturing and industry. The result was the mobilization of the remnant industrial base into the business of automation and productivity science. The act provided massive grants to industry for R&D into even more exotic solutions. This action immediately reversed course and brought productivity back to the US as a national mission. Europe and some Asian countries soon followed suit and initiated similar measures, or accelerated their existing initiatives, like Japan, which had started down this path many years prior, thus the race was on.

Likewise, it was decided that production needed to stay in the developed world, thus the mounting economic and political problems could more easily be resolved. So gradually, from the perspective of public perception, the competitive advantage of shipping manufacturing and industries wholesale to the developing world began to wane, and reverse. Separate bills were drafted and signed into law that ended short sighted trade policies which disadvantaged the US, but were fashioned more like those of Asian countries, thus giving the nation the room necessary to rebuild. That was really the key to restoration of political and economic stability, particularly in the US. Indeed, everywhere things began to improve. The exception is what resulted in the underdeveloped world which suffered a breakdown of social order, due to the contraction and reversal of their industries, without alternative for their now underemployed people, but that was no longer the concern of the west.

Despite the political risks associated with the chronic unemployment that persisted, and was likely to increase, as a result of the productivity dislocation brought on by technology, it was finally, unquestionably demonstrated, that massive increases in automation could have an effect in reversing the impending economic steamroller, by keeping the production in the US instead of offshore. This kept the gain and wealth generated inside the country, resulting in manufacturing and productive parity with the developing world and their cheap labor advantage, which allowed for a generally good economic climate in the developed world.

Wide Commercial Adaptation

[MD Narrative]: Some ballyhooed that the new direction created by the act was protectionist and unfair in trade, and that it would have disastrous effects on the economy and would lead to wars; although it was not certain how things could have gotten any worse, particularly due to reckless Keynesian spending and globalist trade policies pushed by elitists, which had preceded the impending disaster by trading away our manufacturing and industrial base and our marketplace, then ceding our sovereignty to supra-national arbiters of the global economic pie. Indeed there were some dislocation of workers, but the nation was soon on a corrective path and began to recover very strongly, and the new course of things was well along the way. Within a decade, companies like Dalbots, founded by my father Boirix, were producing humanoid machines that could do the exact same work as low skilled workers, and others were doing diligent research to unlock the secrets of the human brain: speech, perception, cognizance, and inductive reasoning, and work in physiology, and bio-mechanics, and the man machine interface; in order to produce machines close to humans in many respects, able to interact with and learn from them, to facilitate and enhance human productivity, certainly where it comes to physical or mental work.

§

Despite the detractors of its safety and efficacy, and the fact that accelerated adoption of automation technology that was intentionally used to displace human labor, was considered by some to be outrageous, it was quickly and decidedly embraced by business and industry. Indeed, virtually every area of commerce was affected. Most would say that it had a greater impact than that of the advent of the automobile, the telephone, the electric lighting of the nation, and the computer and Internet combined.

Eventually, the ability for a company to start up and be profitable with a significant reduction of the past usual risk of failure became a reality, and the economic impact was felt far and wide, as hundreds of thousands, maybe millions of new small and medium sized companies sprouted to profitability aided by new forms of automated machinery, while the more established large corporations and multinationals experienced unprecedented growth in production and profit. The world began to change and the rules of commerce and business were modified. Entrepreneurial success in business became considerably more accessible to the little man, who, was now less compelled to sell a majority stake in his enterprise to large pools of capital, just to keep it alive.

Adam and Woman

And the Lord God said, It is not good that the man should be alone; I will make an help meet for him...

...and the Lord God caused a deep sleep to fall upon Adam, and he slept: and he took one of his ribs, and closed up the flesh instead thereof;

And the rib, which the Lord God had taken from man, made he a woman, and brought her unto the man.

And Adam said, This is now bone of my bones, and flesh of my flesh: she shall be called Woman, because she was taken out of Man.

Therefore shall a man leave his father and his mother, and shall cleave unto his wife: and they shall be one flesh.

And they were both naked, the man and his wife, and were not ashamed.

CHAPTER FIVE

The Machines

Extending Human Labor

[MD Narrative]: What had taken place in generations past, when it was discovered what the computer could do in terms of extending the dynamic working range of the human brain, particularly when it was made human-portable, and could enhance the management and flow of information, extend memory, and ease every day tasks, was re-imagined, except now it was the same concept and effect applied to the human body, which greatly enhanced the human labor yield. As in the past when small computation devices extended the function of the human brain, now it was full size humanoid machines extending the labor function of the human body.

Regardless of the intent of the act, it set in motion the quest for artificially intelligent machines and equipment that could act as aides to industry and commerce, and as some predicted, it quickly became an outright replacement for the common labor worker. It is accepted that this dawned somewhat quickly in about the year 2035, and there were significant numbers of labor jobs threatened, such as agricultural, food preparation, retail services, construction, services like retail banking, hospitality, office work, insurance, financial services and virtually all of the low skilled service trades. These all utilized the productive resources created through the act, and started to replace the humans which had previously done these tasks.

SmartFast

[MD Narrative]: It all started with the advent of the computer and the

microprocessor in the 1960s and '70s. These devices grew in complexity and performance at astronomical rates since their inception. One account had the power of the computer doubling every 18 months, then with the advent of Quantum Influenced Computing Devices - QICDs, and full Quantum Computing Devices - QCDs, Tertiary processor units, Neural Computing and Neural Networking, and Massively Parallel Computational elements, that pace accelerated significantly. By 2040 it was announced that the state of the art had achieved computational power equal to that of the human brain. By that time, the machines employing that computational power had the mental acuity and motor function similar to that of the human brain and cerebral cortex, but could perform physical labor functions and abilities that far exceeded what humans could physically do, such as the ability to perform the same repetitive process exactly the same each time without rest, and without error, and the ability to analyze huge amounts of complex information in large matrices with numerous variables, fast and accurately, thus greatly enhancing the range, flexibility, and adaptability of the work where they are employed.

Shiawasena Machine Co LTD

[MD Narrative]: Development of various components of intelligent machines proceeded by many research organizations and corporations around the world, mostly in the US, Britain, France, Germany, and specifically Japan early on. Most agreed the Japanese were probably ahead in the early part of the development, but due to the Japanese rigid and top-down, highly vertically integrated approach, they were not nearly as innovative in the long run, and because they predictably tended to proceed on a typically large corporate scale. However there were exceptions.

§

The problem with the Japanese approach is that the entire industry had to be invented from the bottom up, and the Japanese did not excel at that in comparison to their European and American counterparts, who had invented the entirety of the modern techno-industrial world in the 500 years prior to that period of time. The Asians, and in particular, the Japanese were wizards when it came to manufacturing and producing the resultant technology, and early in the game they produced some very stout machines indeed, but because of their corporate approach to development, their machines lacked the intricacies and nuances of intelligence that results from refinement by hundreds of thousands of hungry developers worldwide, trying to get a foothold, which was the western approach, and was needed for them to be truly integrated into mass societal structures, and they were very, very expensive because the top-down approach tended to make a one size fits all machine that was not easily adaptable, and virtually required a PhD in physics and mechanical engineering to maintain.

幸せな

マシン株式会社

株式会社

Osaka Japan, Circa 2028. The manufacturing plant is a nondescript and unassuming office complex in the warehouse district of a meticulously clean neighborhood in the industrial center of Osaka. Just a simple sign in Japanese in the front of the building describes what they do; Strong-Happy Machine Company; pretty telling. Inside the main building, the work and manufacturing assembly area is large and clean with florescent lighting hanging from large roof trusses that span the width of the building about 150 foot or so. The ceiling is about 40 feet up, and there are large scaffold type shelving with various large wooden shipping boxes along the sides of the walls giving the impression of a warehouse. Numerous workers are seen rushing about wearing white coats, and hard hats.

Along one side of the building there is a complex of office rooms with standard 10 foot ceilings, and inside is a matrix of modest office furniture, separating cubicles and workspaces. Through a large set of double doors, in the inspection and quality assurance area, workers in this mid sized manufacturing company have been installing a drive system into a humanoid robotic skeleton that looks distinctively like a machine that has the form of a human with 2 arms, and a head, but no torso, just a pneumatically operated telescoping pole about 5 feet high which holds the mass of the machine off the ground, and is able to elevate it up or down to the desired level, attached to a large flat round mounting base. The machine has no skin or covering, but does have what look like eyes that rotate as expected, in a head that can obviously swivel from side to side as well as up and down. The arms are mounted on an assembly which can rotate in a horizontal plane, in circular fashion about 180 degree in either direction, ensuring that the appendages can reach any position within the radius of reach of the arms, which are of considerable length. The machine is an obvious tool for performing tasks with arms and hands.

As one walks through the front reception area of the main building, and down the hallway into the massive warehouse area, there are an assortment of pictures of the various versions of the machines which the company manufactures lining the walls in a historic and chronological order, with the most recent additions being the largest, arranged up at the front of the procession. On the opposite wall of the hallway, in like fashion, a series of pictures in procession of the founders and former presidents and senior executives of the company are displayed proudly. Some of the machines are like the one described, but may have up to 4 arm appendages on the rotating platform, where the presently observed unit only has the 2. Some of the units in the pictures have 2 sets of stereoscopic assemblies mounted on both sides of the head, this way it can see anything in its radius of reach. Some are mounted on powered mobile platforms, some with tracks, and some with tires.

A senior inspector walks into the room where the other 5 team members have just finished re-installing parts of the drive control system which had been removed to satisfy the senior inspector. The work area is strewn about with various hand tools, computer screens and some power tools. Along one side of the room, there are numerous large storage utility cabinets and various pieces of equipment, some mounted on moveable platforms and some portable. The senior workman types commands into a computer terminal nearby and turns around to look at the machine. The machine moves as if it is powering up and initializing. It seems to look and find all the faces in the room as the head and the functional analog of eyes composed of a stereoscopic camera assembly, scans the room. It is an impressive sight. The machine seems to tower over the men to a height of about 8 foot. The senior workman addresses the machine and asks a question of it. The machine then immediately directs its attention to the man asking the question.

"Anata wa watashi o kiku koto wa dekimasu ka?" he asks, which in english is, "Can you hear me?".

This gets the machine's attention, and it replies immediately,

"Watashi wa anata no meirei no junbi ga dekite imasu", which translates to, "I am ready for your instruction!".

It pronounces the words in a loud and deep booming voice. The workman then again commands the machine, "Shite suihei ni hiji de sore o magete, anata no chūshin-bu ni migite o idō shite migite o idō shite kudasai", which is to say, "Please move your right arm up to level, then bend it at the elbow and move your right hand to your center."

The machine immediately and precisely complies making very smooth and fluid motions as it moves its arm as commanded. The senior inspector turns back around to the computer and types in another command. The machine immediately goes through an obvious power down sequence and sinks into a dead repose. He then says to the other 2 men present, "Kono yunitto wa, kurippu o hoji fusoku shite iru dengen basuhānesu o nozoite, subete no kensa o watashimasu. Shūsei wa shite inbentori ni irete.", which is to say, "This unit passes all inspection, except for the missing power bus harness retaining clip. Fix it then put it into the inventory."

He picks up the inspection report attached to the unit by wire, and writes a note, then pulls out a stamp and presses it against the form to pronounce the unit acceptable. As he is doing this, 3 of the other workmen go across the room to various locations to retrieve supplies and tools and the missing retaining clip, then each returns, and as if in a choreographed sequence, each proceeds to perform his very specific part of the task of installing the missing clip along the lower spine section where the power wiring harness runs.

Three other workmen come into the room driving a small mobile hoist, and proceed to disconnect all wiring and related paraphernalia from the unit, then attach a cable assembly from the hoist to 3 positions on the unit. The operator of the hoist pushes a series of levers which lifts the unit off the place where it is resting, and drives it horizontally in a sweeping arc over to the hoist's platform, lowering it into place. They then proceed out of the room with the unit, through the double doors, down the hall and out into the open warehouse area. They drive the unit across the warehouse open floor and into a back area where they place it into the repository.

§

The ride is rough, as the truck bounces down the small narrow brick paved street and arrives at the gates of the complex. A guard waves them through, and they proceed to the back of the main building which is a massive single level industrial building. The truck pulls up to the loading dock and the driver steps out with a handheld inventory order tracking device. He uses it to remotely engage the receptacle on the outside of the docking bay, and the door suddenly opens. Within a few seconds, a forklift arrives and proceeds to drive into the truck to retrieve the unit inside. The driver then engages in small conversation with the senior attendant at the loading dock, and then makes a come hither gesture to his younger deliveryman partner who has also just emerged from within the truck. He makes it up the 6 or so steps to the platform where his partner is standing and the 2 men walk over to the front of the warehouse area, through a set of swinging double doors, down a 12 foot wide hallway.

The hallway is long, about 100 foot in length, and on one side it is lined with windows from the mid torso level up to give total visibility of the processing floor. The younger man seems fascinated as he observes the Shiawasena machines in action like the one they are delivering, which are arrayed on the other side of the glass, so he stops several times along the way just to watch for a few seconds before moving on, while the more senior man seemingly a little irritated, continues to call out to him, "Koji", and gesture as if to say come on now.

In the back of the room, as they near the mid distance of the hallway, there is a queue with a parade of farm raised hogs and pigs being herded through, while a few workers are busy using prods and poles to make sure they continue their appointed walk in the desired direction, which is toward the inevitable. Some seem to realize that there is something threatening ahead of them as they let out extraordinary shrieks of fright, and attempt to flee back in the direction from which they came. Despite the depressing and somewhat terrifying scene observed by the younger delivery man Koji, he is nonetheless fascinated as he observes one Shiawasena unit at work. The unit reaches down into the frantic mass, its head and eyes busy scanning and analyzing the scene for the right pig to process. It then extends the 1st of its 4 powerful arms into the squealing mass and lifts a 300 pound hog up by its hind leg and rapidly swings it over to the refuse area. With another arm, it quickly dispatches the hog with a quick puncture to the head. Immediately the hog ceases all struggling, and hangs still, obviously dead. A 3rd and 4th arm and hand appendage then proceed to assist the 2nd in processing the remainder of the hog, quickly removing the innards, then giving the animal a quick disinfecting hot shower, and then precision butchering the carcass into the various cuts and portions, each of which is flung into a nearby bin by the machine, where the parts are moved by a conveyer track, and further processed in an assembly fashion by other Shiawasena units. In its entirety, the processing of the animal took 90 seconds.

§

[MD Narrative]: In the US, just as in the past with advancing technology starting to proliferate into industry, there were innovations that gave access, and allowed smaller developers, such as the average 'Tinker Joe', or small technology garage start-ups, to make improvements and become developers in the technology, instead of being completely dominated by the rigid, hard, cold and non-functional crap handed down by the vertically integrated corporate world.

The approach of developing 'open source' technology had been pioneered in the US by companies which recognized that good products and good ideas come from many sources and ordinary people, most of whom could not be easily harnessed or controlled, and were elusive because the ones able to do that kind of creative work effectively, coming-up with something great, which really added significant value to the developing technology or platform, were not the type of people that would easily acquiesce and be harnessed or tied down to work in a rigid environment. Generally, the creative and innovative types were not inclined to the restraints and demand of boss, and a job. They were more the free spirit type, analogized by the elusive, beautiful, and fascinating humming bird. As beautiful as it is, and as much as one may wish to capture it and put it in a cage to enjoy whenever they want, it is far too delicate, and would fade, and die very quickly if made captive. So, when creative people came up with a great value added or complementary products, accessories, or services, most the time, their companies would be a target for acquisition by a bigger player to become part of a larger snowball of acquired technology as it rolls down a hill.

The Autonomous Economy I - The Strong Man

Penser et de Faire

[MD Narrative]: The more natural approach to technology development, did not always result in success, because sometimes the acquiring company would find out too late that when they acquired the small company, the genius that created whatever there was of value did not want to come along and be tied down, so they would only stay with the bigger company until they collected the cash, and then bolt for the door to go and do it again. Then the larger company was left with the task of finding someone capable of filling the void left by the originator of the value they sought to acquire; a not so easy proposition. The results were often very dubious, but overall that practice was significantly better than what was happening under other cultures elsewhere in the world. The european approach on the other hand, was very good scientifically and by engineering standards, however, not a panacea.

§

Grenoble France; The name Penser et de Faire means "Thinking and Doing" in English. Alain Michaud is busy running test on the latest pre-release of his company's cutting edge motor control software. Motor control, referring not to the control of an electro-mechanical device with a rotating shaft, but to the group of functions generated by the cerebral cortex in the brain, which in humans and creatures alike is responsible for the ability to walk, jump, run, play tennis, etc. Alain is with his team consisting of 2 engineers, a mechanical technician, and 2 of the business staff in a small room in the company's research center in Grenoble. They are in the meat of an in-depth demonstration of the ability of the newest release (version 2.7), of the Penser software dubbed JumpStart™, and in deep discussion with an investor group from the mid-east that is facilitated by an American from a New York investment group. Bill Lattimore is the scientific and technical liaison for the investment group, and he is engaged in a technical discussion with Alain. Bill has been there before and had some discussions with Alain in the pre-investment due diligence phase of brokering a deal between Penser and the men with the money. This is the 3rd visit for Bill.

In the same room as the men, a section of a humanoid machine is tethered to a long set of cables and assorted peripheral support lines while walking normally on a treadmill as a human would. The machine has 2 legs and 2 arms, and is built with cutting edge bone-like structural members of an engineered metallic material, with joints in those members as those of an average human male in proportion and size. In the same room, there are 2 other test facilities similar to this one, but with mechanical versions of animals; one a dog and the other a large bird like an eagle or condor. The dog stands average height for a family pet at about 2 feet high, while the bird is very large with an equivalent wingspan of 9 feet, which is not unheard of but rare for a real bird, and certainly not typical or representative of what might ordinarily be encountered.

The group has brought an Arabic interpreter, but for the most part he is silent because these men are not attempting to understand the details of the technology but are more interested in the financial aspects such as: what is the status of the newest American and European patents? When will the release be ready to ship? What are the gross sales revenues, and who are the end customers that will buy this product, and how will it be charged; in a license or royalty, etc?

Alain, in his broken and heavily accented english says to Bill, "As you will recall Bill from the last time you visited, we have been building the framework of the system, and we now have expanded the functionality and added several new extensions", a smiling Alain pauses and looks at the group of 2 men dressed in Arab garb, along with Bill Lattimore, as if to let them ask any questions they might have. Bill looks back at Alain, "Lets move on please. Why don't you give a brief overview of the product?"

Alain continues, "We have specialized in balance, equilibrium, and coordination; that is, improving the ability of machines to mimic what humans and animals do; to walk, run and perform routine tasks that humans and animals take for granted. The system uh..., utilize the..., the..., uh..., two hundred, fifty six, Q-bit ViaSec Quantum processor for vector computation algorithm..., uh..., ViaSec processor..., uh..., is manufacture in Quebec Canada. Our software takes a massive, parallel matrix-stream of input motion, parameterized data, like the F thirty two joint vector...; watch the monitor." Alain points to direct the attention of the group to the computer sitting on the table next to the platform.

On the screen is a series of data in a matrix which streams across it as the machine walks. Alain types into the keyboard, and a graph paints the screen which appears to track only one of the data streams. 'F-32' appears in large letters across the top, and the shape of the graph seems to follow a repeating pattern that obviously follows the movement of the machine on the treadmill. Alain then touches the screen and switches the view to one which seems to follow the same data, but graphed in a different way. At the bottom of the screen there are smaller versions of the 2 graphs, along with a 3rd which reveals the other 2 composited together. The 3rd is a 3 dimensional representation of the other two, which show the data in only 2 dimensions.

Alain explains while pointing to areas on the monitor, "This graph is a vector of the F thirty two joint in two axis."

Touching the screen, he switches the image, "…and this is the same with one axis different, and one common."

Again he Touches the screen which changes the image and he explains, "This one represents the three axis together and makes a three dimensional view. It shows a time-history representation of the three dimensional spacial position of the joint over a period of sixty seconds. There are two hundred twenty three Clough actuators, and seventy two three axis rotating position sensors on this test unit."

Alain is a very expressive man and uses his hands often to express his thoughts. He continues while making many hand gestures in the process, "What the JumpStart software does, is take-in the three spacial dimensions for each joint and each actuator in a realtime data stream."

He pauses and steps back, leaning half-reclined, on the edge of a table. He takes a breath and continues, "The data is processed in such a fashion, that real-time target control estimations can be computed and used in a greatly reduced matrix of output vector correction signals. This reduces from the cache, the number of commands required for movement, which increases accuracy and reduces latency. These signals can be taken from, and fed directly to, distributed vector processors in the members, or to a single main controller, or both. Imagine if your own brain could only make one command at a time to move. It would take you forever to go across the room, but in actuality, your brain sends many signals at the same time to your muscles in a coordinated sequence. That is why walking is normal for you. It is the same with this system."

Continuing, "The software uses a special patented algorithm, and produces very large, parallel, realtime output instead of small serial blocks, which reduces latency and jitter, and greatly reduces the amount of data, which significantly improves throughput performance. Any question?"

The group remains silent so he continues, "That makes for very fast, accurate, and very smooth and graceful movements. Much more natural than what results from using a direct control approach."

Bill then says, "Thank you Alain, very informative, and exciting. Can we go into your conference room to discuss business details?"

Alain replies, "Before that, let me show you a demonstration of JumpStart."

Alain walks over to the area where the mechanical dog is tethered, and the group follows him. Three other men as well as the Arabic interpreter walk behind him to the same area while the other members of Alain's staff stay seated on chairs in the background, just behind where the group of men are standing.

The area is about 15 foot in length, and 8 foot in depth, situated next to a drab white wall. There is visible, the rubber track of a stopped treadmill embedded into the floor of the platform which the mechanical dog is sitting in. There is a bundle of electrical cables along with a compressed air hose attached to the model, connected with 6 excess service loop lengths of about 3 feet each, hanging down in successive order, attached by metal retaining rings that slide back and forth over a small thin metal cable suspended horizontally about 7 foot above the floor and fixed at both ends, so as to allow the tether to go wherever the dog goes without hindering its movements.

Alain then starts the direct control, or generic control system software that initializes the mechanical mutt to sit-up and look up, wagging it's tail similar to a what a real dog would do if at attention while waiting for his master to give him a treat, or to command him to do a trick.

Alain then says, "I have started my uh..., competition. It is a standard, uh...., generic..., control system, and the chien is at an initial position waiting for instruction. I shall now initiate a fetch sequence."

Alain types into the control computer sitting on the table next to the test setup. With his attention fixed decidedly on a new task, and with an annoyed look on his face, he is looking about as if to find a lost object.

He then yells out in French to his technician "Thierry, où est le chapeau qui était ici la dernière fois?", which is to say, "Thierry, where did that little hat go that was here the last time?"

Alain is clearly distracted and busy with his eyes darting about, and his head moving while he scans the area for the hat. He opens the door to a large 7 foot utility cabinet standing in a nearby corner, and looks as if he expects to find it. He closes the door.

Thierry then replies, "Je l'ai il à mon bureau. Je vais l'obtenir", which in english means, "I have it in my office. I will get it."

Thierry then walks into the other room, returning 5 seconds later with a small brown french beret hat, which he hands to Alain. "Thank you Thierry", Alain proclaims.

"Now gentlemen, observe", Alain announces, and throws the hat toward the dog.

The dog starts to move to where the hat is, but is hampered and sluggish, walking in a very jerky motion as if in an old stop-action animation. The dog then attempts to pick up the hat in it's mouth. It succeeds after several attempts, then slowly shudders while walking over to the target area designated by an orange circle with a large red X in the center, both of which are painted on the floor. The dog drops the hat on the X and resumes the initial position of sitting with head up.

Alain, typing into the computer keyboard excitedly proclaims, "Now I am going to switch to the JumpStart motor control system." The mechanical mutt goes into an obvious shutdown sequence, and within 15 seconds powers back up in the same initial position as before, sitting, tail wagging and head looking up.

Alain then picks up the hat from the red X and throws it in the direction of the dog, whereupon it leaps up and grabs the moving hat out of the air, as one might observe from a real, young and playful dog that wants to fetch a ball thrown to him. It wags it's head back and forth in rapid succession with the hat held in it's dull plastic teeth, like a dog with a toy. It walks the 5 feet over to the orange circle and red X and drops the hat on the X. It then resumes the initial position again.

The 4 men observing break out into a round of hands clapping, and with smiles on their faces they start to walk out of the room. Thierry announces to Alain, "Il est temps de cesser de fumer patron. Je vais maintenant!", which in english means, "it is quitting time boss, I'm on my way out!" Thierry walks into his office and back out again within 10 seconds wearing his coat, then walks over to where the demonstration had just taken place, picks up his hat, looks at it against the ceiling light, and pokes his finger through a new found hole, then slaps it across his leg, places it on his head, and walks back out of the room.

The 4 men and the translator are in the other room engaged in substantive discussions about what was just demonstrated, while Alain is busy doing most of the talking, attempting to answer their questions. They break and start to wrap up, then as the group walks toward the exit, Bill puts his hand on Alain's shoulder and says to him, "Think about their questions please. I have been in this business a long time, and have grown to recognize the buy signals. I want you to be prepared for moving fast because I believe there may be a buyout or merger offer tendered from the Americans mentioned, financed by this group."

A smiling and excited, but somewhat surprised Alain replies as he grabs Bill's hand tightly with both of his, and shakes once firmly but brief, "I would very much look forward to that. Thank you Bill!"

The Serpent and the Woman

Woman goes to the spring in the midst of the garden to gather fruit and water for both of her and Adam to eat and drink, and she gathers flowers for them to enjoy. She is greeted by the serpent who was there amongst the beasts.

Now the serpent was more subtle than any beast of the field which the Lord God had made. And he said unto the woman, Yea, hath God said, Ye shall not eat of every tree of the garden?

And the woman said unto the serpent, We may eat of the fruit of the trees of the garden:

But of the fruit of the tree which is in the midst of the garden, God hath said, Ye shall not eat of it, neither shall ye touch it, lest ye die.

The Autonomous Economy I - The Strong Man

And the serpent said unto the woman, Ye shall not surely die:

For God doth know that in the day ye eat thereof, then your eyes shall be opened, and ye shall be as gods, knowing good and evil.

And when the woman saw that the tree was good for food, and that it was pleasant to the eyes, and a tree to be desired to make one wise, she took of the fruit thereof, and did eat, and gave also unto her husband with her; and he did eat.

And the eyes of them both were opened, and they knew that they were naked; and they sewed fig leaves together, and made themselves aprons.

CHAPTER SIX

Dalbots

New Helpmate

[MD Narrative]: The early machines had commercially advanced quickly since the 20's due to rapid advancements in the science of artificial intelligence. The application of many advancing technologies had been employed including artificial muscle devices and pneumatic actuators; vast miniaturization of machines and parts; very high density energy storage, fuel cells, and batteries; advanced materials and processes; neural computing and networking, and massively parallel and distributed embedded cell based computing technology.

Quantum Influenced Computing Devices - QICDs, full Quantum Computing Devices - QCDs, and tertiary digital processors, all of which aid machines in learning from their actions, by making estimates, measuring the resultant errors, then estimating and generating the needed correction in subsequent attempts at any task. It was discovered that for the lower functions, this is very much like the way human and animal brains operate, and learn. Using Quantum mechanics based, Massively Parallel, and Neural Computing based systems, it was discovered that machines could analyze a situation, and make decisions in a fashion similar to the way that humans would, given the same or similar circumstances, such as how to safely negotiate a hallway in a house under construction while carrying a large cumbersome load of materials without dumping or bumping; or the ability to diagnose a problem with another machine that has broken down, acquire the needed parts, and perform the repair, then test the machine to ensure that it can safely be re-introduced into it's work.

Dalbots Delegation Tour

[MD Narrative]: By the time of Penser, Europeans had been trapped in the death grip of socialism and confiscatory redistributionist politics for so long that the economic environment was not conducive enough for entrepreneurial endeavors to really flourish, so even though European companies like Penser produced some very excellent technology, for the most part, the talented creative developers in Europe tended to bring their ideas and technology to the US where it could really fulfill its commercial potential.

§

That is precisely why this industry was really born commercially in the US, and as mentioned earlier, although for the most part, manufacturing had been largely moved to Asia in the preceding decades before this point, there had been a change in political philosophy when it was realized that products could be produced anywhere in the world at the same cost because of gains in productivity through industrial automation, and limited political reform, so that trend reversed and brought manufacturing back to the US and other western nations.

Dalbots headquarters, Dallas Texas, USA. Circa 2028. Entering in through the main front doors, and walking up to the large enclosing counter which is the domain of the reception staff, one is reassured by the sweet demure way of the perky young blonde receptionist as she greets visitors with her quintessential and soft, "How ya'all today. Please sign in. Ya'all here to see someone today?"

General Cranket, dressed in the full military garb of a 4 star General, and who is being escorted by the other 3 individuals dressed in civilian clothes, answers and says, "Well miss, we are here on behalf of the United States Government. We need to see old Bo."

Replying, the receptionist asks, "Does Mister Dalgleish know what this is about?"

Bill (William) Dubrowski, leading the procession puts his hand on the General's shoulder and gently walks him about 10 feet away from the others, and says, "General Cranket, please let me handle the liaison. I know you and Mister Dalgleish are friends from way back god knows when, but this is my show."

Standing very upright with his chest heaving, the General replies back saying, "Sir, I will respect your command!"

They then walk back up to the receptionist's station, where the other 2 members are just finishing signing the log, and Bill says to the receptionist, "Miss, would you please inform them that Bill Dubrowski, and the rest of the delegation are here to see Mister Dalgleish?"

A few seconds later, the receptionist says to Bill Dubrowski, "Apparently you are early. Mister Dalgleish will be with you shortly. He is sending Mister Zellhuber to meet you."

Grady Hunt, and Ana (Anastasia) Isidora are also along with General Cranket in the delegation. Peter Zellhuber, then walks from a hallway into the room, and greets the delegation, "Hello, I'm Peter Zellhuber, VP of R&D. I work for Mister Dalgleish. I'd be glad to show you around."

Piping up, General Cranket asks Peter, "When is the big man going to join us?"

Peter replies, "We will join him very shortly. In the mean time, let me show you around. If you will please follow me!"

Peter bowing slightly and gesturing with his hand out, "After you!"

The lobby of the building is 3 floors tall and grand with highly polished granite tiles laid out with an accent border around the room. Modest and simple furnishings are against the walls and extending into the room. The style is distinctive classical-modern, and the walls are host to large array of grand and colorful sculpted artwork, with small modern interpretations of ancient statuary scattered about, but strategically situated. The ambience is warm, elegant and comfortable, and while standing just inside the entrance, 2 foot from the outside door, opposite where the guard station is located, one can not help but notice the large set of mounted Longhorn steer horns with a length of a lariat wrapped in 3 concentric loops around them, hanging just across from the entrance to the lobby; the message is clear; this is Texas, and we do things in the Texas way.

Moving through the hallway, down the corridor, there are a few rooms without windows that are entered through single wide nondescript white doors by employees with identification badges, and key codes to give them access. There are other conference and meeting rooms with windows, where access is direct and not restricted.

Peter explains what they see as they walk down the main corridor, and he keys in a code to one of the doors. He cracks it slightly allowing the delegation to peak in. They observe several individuals seated at small cubicles, each wearing a very thin bonnet, or strap contacting their forehead and face like what one would use to conduct a brain scan, with wires attached. They are all staring at individual monitors which are displaying images.

Peter explains, "These are our conference rooms, and some R&D facilities where we perform some of the magic. The building has 3 floors of about 10,000 square foot on each level. Please turn around and follow me up the stairs."

They walk back and start up a large open central stairway with 2 half switchbacks proceeding up to each level. The perimeter of the lunch room along one side is rimmed by a 4 foot glass wall, and opens up to allow a view of the entrance, central stairway, and open lobby below.

Reaching the 1st landing Peter gestures to the left and says, "This is the mezzanine level and the lunch room. Can I get you all anything to drink?" The delegation look at each other and gesture that they are fine, nothing now. Then continuing up the other switchback to the 2nd level mezzanine, where the small cafe and lunch room are located, with nice new tables and chairs situated, they pause then walk across the floor to the backside of the mezzanine, then look down to the open space on the 1st level. Immediately below where they stand at the opposite end of the main corridor where they entered the building, there is an open room filled with cubicles for employees who are busy milling about.

Across the open stairway on the same level is a wall along one side bisecting the area, extending from the lobby to the back, with windows that allow a view similar to that of the lunch area, with an open area and a 4 foot glass wall about half as large as that of the lunch area situated directly across from there.

Peter gestures to the right side, "Over there, is where the management offices are located. Please follow me up again.", he asks as they proceed back to the stairway.

Third Floor

The 3rd floor is very restricted, and as they make their way up the final level of the stairs, they are met at the landing by a guard, and electronically keyed door.

They approach the guard station, and Peter says, "Mister Dalgleish assures me that you all have been cleared, but you will need to sign in."

General Cranket gets right up in Peter's face, and looks him right in the eye, then he pipes up and with a little bit of sarcastic skepticism in his voice, he says, "Kind of simple for security, wouldn't you say?"

"I can assure you, it is more than adequate. This is the R&D area, and only employees with high level clearance are allowed entry.", Peter asserts, as the group signs the log, one after the other.

As they enter the door, Boirix Dalgleish, and James Bowen are there and both greet them graciously with a large smile. Boirix is especially delighted at seeing General Cranket. "Gerry, how good to see you again…, welcome and thank you for coming!" General Cranket replies, "Likewise Bo, its been a while. Bo, let me introduce you here. This is Bill Dubrowski, Grady Hunt, and Ann Isidora."

"Gentlemen…, Ma'am", replies Boirix, as he gently shakes their hands.

Continuing, Boirix introduces some of his staff, "Gentlemen, this is doctor James Bowen, our director of research, and this is my son Michael. He's a specialist in the security and protocols development, and you have already met Peter who works along with Jim and Michael, also a specialist in security and protocols. If you will all follow me?"

They all exchange simple greetings with Jim Bowen and Michael and proceed to follow Boirix down the corridor to the back where the large room is. Inside the main door is another corridor with rooms on either side. The room at the back of the corridor spans the entire width of the building, and inside are several stations, setup with several robotic machines in various states of assembly. All of them are being used as test and development rigs, so in 2 cases all that can be observed is a head in various stages of assembly.

Approaching one of the station areas, Boirix explains, "With the humanoid robots under development by Dalbots, there are no pretenses about what they are, and what they are not. They are computerized mechanical machines consisting of fabricated parts of metal, rubber and plastics, with electronically controlled, magnetic, lead screw, or pneumatically driven actuators, and motors that make things move about, directed by electronic signals, generated by numerous embedded micro-controllers. What they are not, are actual flesh and blood human body parts, so they do not resemble them in detail, purposefully and by design."

Although, there are arguments raging about whether these machines should mimic humans in their appearance so as to not make human individuals uneasy at seeing them, nor frightened of interacting with them, or whether they should appear distinctive from humans, and in particular, unmistakably a manufactured mechanical devise or machine, much like the automobile or computer.

The station has a scientist sitting at a computer interacting with, and controlling the test subject, which is a partially assembled robotic head. Boirix asks the man sitting at the station, "Please run the expression sequence?"

The man then types some commands into the keyboard, and the head begins to make some expressive movements of the eyes and mouth. A series of emotional expressions, smiling, listening, pouting, etc.

Boirix continuing, "Dalbots decided to take the approach that appearance was secondary due to the detail materials and labor costs, and because there were of necessity, going to be demand for models that from an application standpoint, could not be a visual analog of a human."

It may have extra arms or joints in extremities, or tools where hands should be, or other such aberrations from the human form. Research was unequivocal, so it was decided, that instead, employing designers that could add aesthetically pleasing and non threatening features into the parts that make up the body and appearance, and allow interacting human observers to not mistake it's form in any way, would better serve the interests of all. The human psyche to machine interface clearly became paramount in the consideration of all the tradeoffs. Much research was done to effectively design these features. Consideration for these features were such as, how the eyes were constructed and how they moved? Was a functional mouth needed, to what degree, and what purpose was served by it? Did the unit have visual ears, and did they play a psychological role in non-auditory communications with humans, etc?

Boirix continuing, and gesturing at the man running the test, "Human psychologists, including our own Dr. Sreenivasan here are employed to help design and test these features, because the intent is to make the units appear friendly and helpful and not threatening. The eyes were analogous to those of humans because, it was shown that the eyes in particular, and surrounds, are the most expressive visual of the human head, and much that is non verbal is communicated by humans with the eyes, followed by the mouth. Doctor Sreenivasan, would you like to say a few things to the delegation?"

Dr. Sreenivasan then stands up and says in his distinct punjabi accented english, "We have learned that the eyes are most important due to the way one human infers things from another's eyes, and not because the person communicating is consciously attempting to send a message using the eyes. Mostly, it occurs without any knowledge, or express intent of either person, but the communication is more on the sub-conscious level. The same with the mouth; one may infer a tremendous amount from another's emotional state and intent by a smile, the brightness of the eyes and position of the eyebrows."

Dr. Sreenivasan then types something on the keyboard of the computer, and gestures, pointing at the partially assembled head component. The head then changes the expression on the face, first forming a frown with an expression of unhappiness, then a smile. He then says, "If the eyebrows are forced down in the center, the apertures of the eyes are small, and the orbits of the eye socket is reduced, that may be interpreted as hostility, while the eyebrows raised and a smile on the face, usually conveys pleasantness, and a happy mood."

Dr. Bowen then chimes in, "Thank you Doctor Sreenivasan", then continues, "The eye analogs for the Dalbots machines are assemblies of stereoscopic cameras with variable apertures mounted in orbs, so as to allow the visual interpretation of rotation up and down and side to side, like that of a human. Being an opto-mechanical image device, it may possess other features that humans do not, such as the ability to view things in other than visible light, like infrared and ultraviolet, and the ability to switch optical filters in or out, like a human may do with sunglasses. Other features are, the ability to see brilliantly in very low light, and some are features based on the old fashioned digital camera developed decades before, such as digital image or motion stabilization, with both optical and digital zoom, with factors as high as fifty X. The basic detection devises are independent and large, about twenty two megapixel Charge Coupled Devices whose outputs are fed to the main visual processor, which can support up to six such devices. This gives the models an approximate five times better vision than humans in terms of view distance, resolution, and the ability to see in extreme high and extremely low level light; and the ability to see in non human viewable light."

Dr. Bowen then looks at General Cranket and continues, "Gentlemen, these machines can stare directly at the sun, and can see perfectly in virtual darkness, and generate images in the visible as well as UV and infrared spectrums. They can switch in optical filters in combinations that allow only select narrow bands in, or filter out only the same narrow band at virtually any level or wavelength. They can record video streams in full stereoscopic, at full resolution, with audio, and other vital sensor data, creating a permanent record of their daily activities, and that of everyone they come into contact with, similar to the airline black boxes we've all heard about."

The group follows Dr. Bowen as he proceeds to another station which has another head and a partial torso, also in an incomplete state of assembly. He asks, "Now can anyone tell me; where would you normally expect to find the brain? Anyone?" No one answers so he continues.

Picking up and holding the partially assembled head unit he answers, "If your answer is the head, you'd be wrong. Robot brains are in a totally unexpected place. The brain, a cognizance engine of sorts, is a small four by five inch assembly which parses all sensory input, sorting it for cognizance processing, and it is located in the lower midsection, the same as it is with some humans. Sorry couldn't resist, just a little robot humor."

He continues, "Can anyone tell me what cognizance is?", again no response so he answers.

"Cognizance is the output or result of what our brains do when they process sensory input. It is our perception; our awareness of ourselves and the world around us. Although it is a much lower level and much less complex in these machines than the same in humans."

He continues, "As with the vision center, similar technological enhancements also apply to the machine's analog for ears. There are generally two, positioned on either side of the head as with humans, enhancing aesthetics, and they are very sensitive microphones fed into the audio detection systems, which utilizes digital signal processors to perform automatic level control, audio compression, speech processing, and any other audio processing requirements that may arise. Just as with the eyes, the signals produced are digital, and fed to the auditory processor inputs via serial interface."

Dr. Bowen, then gestures to the mouth area of the test machines, and continues, "The analog for the human mouth is constructed using 2 assemblies consisting of several segments of inflatable tubes, arranged concentrically in 3 linear groups about a central rubber core. One assembly is for the top lip and other for the bottom. These assemblies change their shape based on the degree to which each of the inflatable tubes is pressurized. The units can convey auditory information through a speaker and audio system, however there is no attempt for the lips to follow the patterns and shapes made by the human mouth while speaking. Instead, the shape of the machine's mouth analog, is simply used to simulate and convey an emotional state, or mood to humans. For example, a smile shape to convey all is ok, and a frown shape to convey confusion or angst." Dr. Bowen continues, "The speech subsystem is composed of a core processing module for parsing and interpreting human input speech, and synthesis of the same for output, interfaced to the multichannel audio processing subsystem, including, power amplifier, mixer and crossbar. Sound is conveyed through multiple channels to high quality industrial speakers, which resist the destructive effects of weather and harsh environments, or audio may be directed through other interfaces."

Continuing, "All of the various subsystem processors are distributed throughout the machine, with some in the head, but as I mentioned a moment ago, mostly the torso and various other places, while the extremities contain their own processors."

Dr. Bowen then directs the group over to a newly developed

demonstration station for visitors and says, "Gentlemen shake hands with the ARCHIE!"

A hand and arm assembly attached to a station is set up so that whenever someone puts their arm into the sphere of attention or space monitored by the machine controlling the mechanical arm and hand, it senses the human hand and reaches over to grab and shake it, which startles many people who are surprised and giggle by the experience. Dr. Bowen explains the Penser et de Faire system, the actuators, and the tactile sensors in the hand, which all play a role in making the machine gentle as it grabs and shakes hands with the visitors, who express amazement and comment about how gentle and natural it feels.

Dr. Bowen asks the group, "Did you know robots have twelve fingers, and are ambidextrous in both the right and left hands? They have four fingers with two opposing thumbs on each hand, one on each side, which may feel a little strange, so, ma'am, it may startle you. If you are right or left handed, it doesn't matter; it senses which hand you use and will shake yours with a completely natural feel. A fully assembled unit can shake your right hand with its left and vise versa, for a completely natural feel, due to a symmetric bi-directional clasp. Go ahead and shake hands with ARCHIE."

Ana Isidora blushes a little as the group lines up one after the other to engage the mechanical arm and hand. As long as the person's grip remains firm, the robot hand will hold the grasp, but as soon as the person relaxes their grip, the machine lets go. The machine will very gently shake the hand up and down initially, but as soon as it senses the human stopping the gesture, it stops as well. In this way, one is impressed by the gentle nature of machine behavior. The arm and hand assembly is programmed to recognize a myriad of hand shake protocols including the gentlemen's handshake, hand-bump, inside finger-palm tickle, double reciprocating palm-swipe, high-five, and the half-grasp granny-shake. The group stays for a few moments as Dr. Bowen explains then demonstrates a few of them, and the group tries them out.

Dr. Bowen continues with the tour, "The chassis is made from engineered high impact plastics, extremely light metallic alloys, and composite fiber to make the unit light, very light in fact, but retain approximately four times the tensile and compressive strength of their representative human body constituent parts."

Continuing, "Movement is powered predominantly by compressed air with a compliment of redundant tanks that can handle up to 1000 PSI pressurization. The compressed air is used to power Clough actuators, which function as air driven muscles, but deliver up to four times the force of a comparable size biological muscle. A network of heat detection, and tactile sensors are distributed throughout the appropriate surfaces, such as fingers, extremity joints and the like. Other sensors that can be installed include ambient temperature, humidity, wind speed and direction and other weather related parameters, bio-material, DNA and human or animal blood and body fluids, radio frequency, electro-magnetic and electrostatic fields, micro-vibration, complete chemical analysis capability such as firearm residue or those that may be toxic, smoke; even nuclear radiation. The list is endless and mostly driven by customer requirements."

Dr. Bowen then pauses, and asks, "Are you all still with me? Any question so far?"

The delegation is silent for a few seconds, then Bill Dubrowski asks, "Can we take a break in a while and go into a conference room? I'm sure we will have several questions", to which Boirix replies, "We are close to concluding this portion of the tour and demonstration. We will take a break for lunch shortly, then go downstairs where we can field any questions you may have, please doctor Bowen, continue."

Dr. Bowen then continues, "Some systems are driven electrically, so a complement of a small hydrogen fuel cell, and redundant high capacity rechargeable batteries are carried to supply actuator power and to power the various processing systems. Fuels cells, and both air tanks and batteries can be charged, or changed out quickly for continuous uninterrupted service. Fully charged, you can expect approximately 8 hours of continuous medium duty service before a recharge. Several small micro-gyroscopes are employed for balance, orientation and for seismic and vibration detection."

Then pointing to the front panel on the torso, he continues, "External panels that fit the front or rear of the torso are installed with a small high resolution display and keyboard hidden behind inside. Thats the cheap option, or as you can see, the optional WrapView™ flexible touch-panel display can be applied to surfaces, such as we have here. These displays have a unique feature so that if you feed it a viewing vector, meaning, the angle with which you view the monitor, it can correct the image so you will not see it distorted, even when wrapped around a curved surface. The machine's vision system senses where you are looking with relationship to the display, which generates the viewing vector input."

Continuing, "Data access is via wireless radio frequency wide area network, and local network modem. They are generally designed for continuous connection to the cloud for soft-updates, downloading utilities and Application-Packs, and exchange-loading their daily activity and knowledge-base schedule, and task record. Are there any other questions?"

The group is somewhat overwhelmed by the technology tour and are looking somewhat haggard, so sensing the fatigue, Dr. Bowen decides to cut the tour short, "We didn't get to everything, but I am sensing ya'all are needing a break, so that completes my portion for now. After lunch, I'm sure you will have some questions for our security and behavioral people."

The group walks back toward the first station. Stopping for a moment, they further observe what Dr. Jankto Srinivasan, a phycologist, is conducting, which are emotional response tests. He, and his intern are sitting at a small table that has 6 monitors at a station, displaying images from several cameras setup to watch several live human volunteer test subjects located down on the 1st floor in a room off the main corridor. The test subjects are busy watching a series of images that depict various animated artistic renderings representing robotic faces and bodies, and their various expressions, and gestures, interspersed with extreme beauty and extremely grotesque human faces, bodies, and poses. A baseline response is generated by the extreme images and used to calibrate the response of each test subject. The test subjects are wired with electrode sensors attached to their heads and arms, in order to measure the emotional response at observing each image. The test is designed and administered by Dr. Srinivasan, his assistants, and various interns.

Leaning forward in his chair, Dr. Srinivasan, speaks into a microphone stationed at the desk, and directs the test subject, a petite blonde woman in her mid 30's sitting at a table in front of a monitor, "Please empty your emotions, and any preoccupations you are thinking now. Try to concentrate on the images as they come across the screen. You will need to concentrate for 5 seconds at each image."

Peter Zellhuber then says, "These are the people we saw downstairs in the room earlier."

The group then breaks, and starts walking to the door, and Boirix says, "Gentlemen…, and ma'am, we have made provisions for lunch, if you will follow me down to the lunch room."

Arriving back at the mezzanine level, they proceed to a section in the lunch area roped off and reserved for them. There are 8 places set with tablecloth, and wine glasses. There are a few people dressed as wait staff standing by next to silver serving and warming tableware. Each one seats themselves, and the waitstaff begin to serve them the bounty.

During lunch several more questions came up about the emphasis placed on designing and controlling of the appearance, and psychological impact and perception of the machines being produced.

Ana Isidora asks Boirix, "Mister Dalgleish, I have two questions for you, first, is the name of the company Dalbots from Dallas and Robots, it seems like that could…?", Boirix interjects, "No…, uh…, Dalbots is from my name Dalgleish, and Robots. Dallas is just our home town. Sorry I interrupted you, please continue."

Ana then continues from where she was cut off, "well, and second, can you explain why so much emphasis on designing your machines to prevent them from seeming as a machine? Why do they need to seem like they are…, uh…, almost human, if I can use that description? Why is it so important to make the machines look friendly, and why do individuals need to feel, ummm…, safe…, not afraid, I mean it makes sense to me, but it seems maybe a little bit over wrought?"

She then proceeds to perform and old fashioned and fading tradition which is to pass her government business card to the Dalbots staff, as the rest of them do as well, throwing their cards to the middle of the table. The Dalbots staff likewise throw their cards on the table and pick up each of the cards from the delegation, who reciprocate as well. Some engage the '*gather contacts*' function of their hand held com devices to contact the devices of the others in the room to solicit and receive contact information.

The function operates in the close proximity of another person whereby their com devices will present a request for invitation and contact information, along with their name and mugshot to the other party's device. The receiving party may then, at their leisure, accept or deny the invitation, and request for contact, in which case the 2 devices either further exchange that decision and data, or if rejected the request is destroyed. The contact information is only forwarded if authorized explicitly by the person being requested of. Often people will engage this function on their devices whenever they are scheduled to go into a meeting of strangers. On its own, the devices will engage all others present that have allowed their devices the same function, and then sort out, and authorize the keepers later.

Boirix then proceeds to answer Ms. Isidora's question, "Thats a great question, Miss Isidora.", he is cut off though as Ana Isidora interjects, "Its Miz..., Ana, actually Anastasia Isidora..., its a greek name, my family is greek."

Boirix pauses then cautiously proceeds, "Uh, Miz Isidora, good question. We have determined that it is vitally important, from the standpoint of acceptance in the marketplace that our product give the perception of safety and trust. We want people to accept them as trusted helpers, that will make their lives easier, and not fear them because there have been so many stereotypes perpetuated in the media about them threatening to wipe out the human race, and other such non-sense."

Boirix continues, "We have to counter that perception and fear. There is a part of our product offering we have not discussed yet which we will see after lunch. After you hear about it, I think it will help you understand how we are the best suited of everyone in our industry to address that negative perception. The product offering is call the X-Pack facility, a software offering that will allow organic evolution of our products to a natural symbiosis with humans. Inside the X-Pack facility, we have something, the roots of which extend back into previous generation Dalbots technology, which we have developed in the last 5 or so years, with in-house, and with several 3rd party developers. Its dubbed and trademarked as Fac-Sentient. It's a software facility that gives these machines a virtual human emotional response system. You will see it tomorrow. I'm telling you, you cannot tell that they are not alive. We are very proud of it."

Conference Room 1B

After they conclude lunch, the group goes down to a large conference room on the 1st level to continue the discussion and demonstration. The room has a large table that seats about 20 around the perimeter with as many chairs. There is a small buffet style table against the far wall with several pitchers of water and a few glasses which a few of them grab, then take to their seats. Mid-height on the wall In the front, and backside of the room are large, 70 inch, diagonal glass work boards, and display monitors that have an overlay for images and drawings created via several electronic tablets with stylus sitting on a table nearby. The image tablet allows the user to view, and control everything on the main screen including anything he may download, call-up, initiate, invoke, or draw on the tablet display, which will also let the viewers in the room see the overlaid doodles. The user can also wear a small clip-on microphone allowing their voice to come from the direction of attention.

Boirix stands up in front of the large display in the front of the room and says, "Again, I would like to thank ya'all for coming today. We have a few more things to cover, so I would like Peter Zellhuber to take the floor, and demonstrate some of the features of the X-Pack facility, and in particular, the Fac-Sentient. Then if Michael would come up after Peter, and talk a little about P&D to conclude. I would also like to remind the delegation that we have scheduled a live demonstration of our latest prototype, the ARCHIE 500 tomorrow, with a late afternoon event that is partly demonstration, and partly celebration. Bill, General, can I confirm you will be staying until tomorrow evening?"

General Cranket says, "I'm not going anywhere. I want to see this brute in action."

Bill Dubrowski looks at Ana, and then at Grady, for confirmation, then Grady Hunt nods and mumbles, "Oh yea, I'll be here", and Ana Isidora says, "Me too."

Bill looks back at Boirix, and says, "Guess thats all of us staying." Boirix then says, "Wonderful, were on for the full show. Ok, Peter you have the floor."

Peter then picks up a floor tablet, and sits down with it while attaching the small microphone to his collar and sitting back down.

The Autonomous Economy I - The Strong Man

He had previously brought up his selection of presentation slides while Boirix was speaking, so the first was already shown on the large display in front. He begins by asking each member of the delegation to give a brief description of their particular job, and what they expect from the systems they are looking at, "If each of you would state your name again and what your job is, then please state briefly what you expect to see, and how the Dalbots technology might fit the requirements. Bill if you would start first?"

Bill Dubrowski then stands up in place, and states the case for himself, and the rest of his staff present, which is Ana Isidora, "Well, I'm Bill Dubrowski, Under Secretary for the Office of Science & Technology Directorate, in the Emergency Services Agency of the US Government, Miz Isidora is a special assistant from my office. Ana, would you like to say something?"

She stands up like her boss, "Hello, I'm Miz Ana Isidora, my official title is special assistant to the undersecretary for the office of science & technology directorate, and I'm here to assist Bill Dubrowski in a fact finding assignment for the office."

Ana then looks back at Bill Dubrowski, as if to say, "I have finished and I don't want to take the spotlight away from you sir."

Bill Dubrowski then continues, "Well I guess our mission here is to get a glimpse of what this technology is about, and what it can offer the people of the United States. Our main job here at the Emergency Services Office, is to be first responders in emergencies that threaten the lives and safety of the citizens, and the national security of the country. We have been told that this technology holds the promise of enhancing that role..., uh..., ummm..., and so we are here to evaluate, watch and learn."

Next in line is Grady Hunt who, sitting at the large table starts to speak, "Hello, my name is Grady Hunt, I'm a senior accountant with the United States Congressional Office of Management and Accounting. My office sent me as a special attaché for Senators Schuman's, and Stumble's offices. Mainly I'm here on a fact finding mission to evaluate the technology, and report back. I'm here to evaluate, and examine the feasibility and costs to the government for deploying the technology in various civilian or non-military capacities. Ummm..., I guess I am also a bit curious..., of what these things are capable of since passage of the productivity act, and I know the two senators are eager to evaluate the progress."

Next the General stands and turns to address the people sitting in various places in the room; he paces a bit with his one hand holding the other behind his back, "I'm Jerome Cranket, General, 4 stars, United States Army, and I keep the country safe and secure, at home and abroad.   In my spare time, I mainly break things and kill people…, ha ha…, seriously, me and Bo here go back what, about 10 years would you say? Well anyway, sometimes I wear more than one hat.   Today I am special envoy from the Secretary of Defense, and I work procurement for the Army Mechanized and Mobile units, and the Office of Defense Procurement Services.   The secretary and I, wanted to see what Bo had up his sleeve lately. We have heard a lot, and I wanted a first hand look, and I can promise you that if these boys can do what we've heard they can, we are not here just to kick tires, but we are ready to sign contracts."

Peter then speaks again, "Thank you all for the brief introduction about yourselves. I'm going to talk a little about the sentience aspect of the systems, and give a brief overview of our approach."

Peter is interrupted by Ana who asks, "Mister   uh…, ummm", Peter finishes her thought, "It's zell, huber!"

Ana continues, "Peter, what is meant by senitence, or sentience, whatever the word is? Can you explain it?"

Peter then continues, "I was just getting to that.   Sentience is something in humans described by the popular culture as the essence of being alive. It is the emotions, that we humans, and even some animals, feel, and express whenever the occasion demands. If I am happy, most people can tell because there is probably a smile on my face, and I can tell when Michael is scheming because he looks guilty, just kidding MD.   The machines are not really alive, but we can simulate these expressions in software.   Our patented, and trademarked product is called Fac-Sentient, and it is something that we started pioneering several years back in our kiosk series. When we started implementing machine vision and auditory functions, we found that we could get a better response from people if we engaged the individuals the way another human might.   It turns out that people are more receptive to a machine that can observe you, and watch you, if it has some aspects of humanity in it. I mean, the trust factor is much greater this way. Look at the screen to get an idea. In the clip, you will see a series of human-machine interactions captured live in real service. It illustrates the before and after effect of sentience enhancement."

The large display then plays a 4 minute video of 6 different people, from children to old people, even pets were included as they engage the Dalbots previous generation kiosks at retail outlets, and their expressions as the machines interact with them. At the beginning, the first 3 cameos, there was no attempt to engage the humans, so their expression during the exchange were mostly lifeless and cold. With the last 3, when the predecessor to the Fac-Sentient™ software was installed, the people smiled and talked to the machines more like they were talking to a human, and sales at such machines climbed dramatically.

After the video clip completes, General Cranket is sitting at the table staring intently at Peter with his arms crossed with a somewhat disapproving scowl and look of skepticism on his face. He says, "Well that is all well and fine, but just before we kill them, we are not interested in making friends with them, hell, I know we don't care how they feel about things."

Boirix interjects, "Good point Gerry, but your requirements are not the same as for most. We have something else in mind for the military. Keep in mind that these things will interact daily with the rest of us, and we cannot afford to scare them."

Peter continues, "The software is contained in another new Dalbots innovation called a soft execution wrapper that protects execution, and makes sure there are no security holes in its implementation. We designed the system to be open, meaning, we created an open API, that is, an application programming interface, and special development tools for programmers, that allow third party developers to build software systems for these machines. So them, us, the consumer, everyone is protected by the wrapper. I know I am getting too technical for some of you, but please bear with me, because I am almost finished. The bottom line is, we would not be able to offer third party support if not for supervisory execution wrappers, which third party software is bound within. It would need to hack its way out of the wrapper before it could cause any mischief, and that is a virtual impossibility."

Bill Dubrowski then interrupts, "You say a virtual impossibility, but that implies that there is some very small chance that it could happen. I'm not sure what that would mean if it did happen, but it sounds like that would be a bad thing!"

Peter then replies, "Well, there is always a small statistical probability, but hey life is full of risks."

Peter notices Boirix has a rather negative disapproving scowl on his face, and is making short barely noticeable gestures toward Peter with quick short head shakes as if to say, "You are getting into sensitive territory, move on!"

Peter continues, "I can assure you, we have studied this issue exhaustively, and taken every precaution to assure safety."

Peter pauses, then cautiously continues, "Let me conclude by mentioning another aspect of this innovation, by the way, the wrapper for third party applications, is what we call the X-pack facility. It is patent pending and trademarked by Dalbots. Michael Dalgleish is going to speak next about another layer, or aspect of why things are very safe and secure. He will discuss what we call the P&D for Protocols and Directives facilities, but before that, I would like you all to watch the next little clip of about 5 minutes which will explain the X-pack facility in a way that will make it more clear to the layman. General Cranket, I know you are following most of this aren't you?"

General Cranket smiles and nods a little and comments, "You bet your sweet ass I am, you don't get to be in my position keeping your head in the sand!"

Peter then starts the video clip of about 5 minutes worth of retail marketing fluff, that introduces aspects of the X-Pack™ facility in very simplistic terms. It explains the Kno-Pack™, and Do-Pack™ offerings which are X-Pack™ templates for the implementation of special knowledge and skill, then the Persona-Pack™ template, which offer a framework for adding special personality profile aspects such as language, and accent, as well as quirkiness like a sense of humor, or maybe taking on the persona of a movie actor like James Cagney, or John Wayne. This is followed by a discussion of the Work-Pack™ template which is a general computing environment for those interested in using the machine as an applications platform. The final discussion of the video clip touched on the ability of the machines use of artificial intelligence to learn and acquire cumulative knowledge and skill from its environment, experience, and training, then compile that knowledge or skill using the X-Pack™ facility into a new offering, or improve an existing X-Pack™ already on the market. The latest resulting cumulative X-Pack™, could then be a new or updated offering for license to others, or kept proprietary, and used to expand an existing workforce.

While the video clip was in progress, 2 other men dressed in business attire entered the room, and sat next to the wall waiting until completion, after which they stood up and walked over to where Boirix was seated. Boirix then stood up and introduced them, "Gentlemen…, and ma'am, I would like to introduce you to two other members of the Dalbots team. They are also here to answer any questions you might have. This is Greg Tishman, Dalbots CFO, and Laurence Bartelli, our marketing manager."

Laurence Bartelli then reaches his hand out to General Cranket, and says, "Glad to meet you sir, I'm Larry, Bartelli", he then does the same to the other members of the delegation one at a time, handing them a business card.

Boirix then says, "Bill, Miz Ana, Grady, Laurence is the person that will get you quick answers to any questions you may have after you leave here. Make sure you exchange his contact information." Boirix carefully avoided including the general in that because of their special relationship as buddies going way back, not wanting to seem as if he was pushing the General off on an underling.

Greg Tishman says, "We are sorry we could not be here earlier to meet with you, but we had another engagement with a group from industry this morning and we just got back. That is the fourth such meeting this week, we are really getting some excitement out there with this new product offering."

Boirix then says to the 2 men, "Gentlemen, if you will excuse us now, we have to get back to our presentation!", so the 2 men started toward the door, and Laurence Bartelli says, "Great to meet you all, I am looking forward to hearing from you soon." They walk out and shut the door behind them.

Boirix continues, "If we could all resume our seats again, we still have some ground to cover, and I believe my son Michael was about to cover some aspects of the P&D. Michael and Peter head up this team, and both of them work for Doctor Bowen. Michael is a security specialist with a masters degree in Computer Science…,

Michael!"

Michael says, "Thank you Boirix", as he stands and takes possession of the magic tablet left next to where Peter Zellhuber is sitting. He clips on a microphone and begins, "This is a special group here today, and because of that, I am going to discuss some subject matter that is sensitive, and may not be known by any one that is less than your alls levels of security clearance. If this were a group from private industry, I would cover much of the same material, but I would leave out some of the material related to high level security and authorization, and several other discussions that relate to matters for government only, but since you are of the highest level, for our purposes, everything I cover will be for your eyes only, so to speak, so, I will also very carefully identify those areas that are under the clearance requirements, so that you will recognize where they lay. It will also be very evident as we proceed."

Michael continues, "I am going to talk about Protocols and Directives. A protocol is a procedure framework that governs the behavior of the machine that employs it. A protocol can be viewed as the framework that defines and limits the extent of what can, and can not be done. This should not be confused, or misconstrued with that of an inhibition. An inhibition is something that is employed in protocols, but is negative and self-restraining in that it is usually meant to prevent something from happening, while a protocol can be viewed as either a positive, an affirmative action or allowance, or a negative restrainer, a prohibition. Any questions so far?"

Michael looks around the room and does not immediately see anyone looking like they have a question, nor even a confused look, so he continues, "A directive is the positive component to the protocol. A directive is just what it implies. It is a command or an order if you would like to use that word. General, I'm sure you can relate to what I am saying?"

Michael pauses, and the General nods, but remains silent. Michael continues, "The Dalbots machines starting with the ARCH 500 models, have a specific set of access levels which authorize certain protocols, inhibitors, and directives. Together we call these the P&D for protocols and directives. Why the word inhibition was left out, I do not know. I imagine the marketing people did not like that terminology, so it was left out, anyway, I will cover them briefly."

Michael continues, "Level one is the most basic and guides the retail user, or owner in the process of initializing the machine with parameters such as name, recognized masters, owners, designated masters, designated authority, home-place, passwords and pass-phrases, and so forth."

He continues, "Level two is the next level which allows the retail owner slash master to enable the machine with duty or employment protocol, by the addition of software, using the Dalbots X-Packs which we covered earlier, or to initialize only the standard ones which come packaged."

Continuing, "Level three is the third retail level that facilitates the reseting or re-initialization of the machine. If for example one wanted to wipe its memory, and reinitialize the machine for re-sale."

He continues, "Level four, which we call Nirvana, is the first level for government, so retail users do not have access to this level, nor the knowledge of its existence, so this is where your security level starts. Everything I talk about hereafter is under your security clearance. Once this level is accessed by an authoritative custodial jurisdiction, it can initiate a limited number of machines homed in a specified area, to act in support of law enforcement, as eyes and ears in surveillance operations, in support of criminal investigation, and apprehension, and so on. They can also be massively deployed for emergency services, such as search and rescue, and emergency shelter construction, and so on."

He continues, "Level five, called Olympus, for civilian government, military, or law enforcement, is similar to those of level four, except that it is designed more for large scale disaster and a higher governmental authority level, for federal or state emergency services, such as earthquakes, floods, or for threats to national security, search and rescue, and so forth."

Bill Dubrowski then remarks, "That is more what we are interested in, and what would serve our purposes. Do you actually have this working now? When can this be deployed, sorry for interrupting, its just that I am really astonished that this is being done!"

Michael continues, "Very good questions mister Dubrowski, I'm sure we can get you answers to all of those questions before you leave."

Boirix then interjects, "Bill, why don't you and I talk again after the presentation.   I would like to get Larry Bartelli back into the conversation with the specifics of your questions.   Michael please continue."

Michael continues, "Right Boirix, so then, I was explaining the government level five access, ummm…, the access, again, can be initiated, in a limited number of machines by a custodial government having jurisdiction in the area where the machines are homed. In the case of the federal government, that means anywhere in the united states.   Operational directives, and protocols may be initiated, terminated, or changed in real time by human facility, or may be set for a limited term for autonomous operation, but only if initiated by a human having jurisdictional authority."

Michael continues, "There is one more level, level six, but I'm afraid I can only discuss that one with the General, so now would be the time for questions if you have any."

Ana Isidora, then asks, "Now let me get this…, when the government wants to use these machines, where do they come from? Do they come from people that own them or what, and what happens to the people that loose their property, does it come back to them later…, I…, uh…, Don't?"

Michael answers, "a very good question Miz Isidora. There is part of the protocol selection process that limits the number of machines to be called into service. It is based on geographical and other criteria gathered by the network resources, and will carefully limit impact on the owners, some of whom may be very much in need of their machines when a greater emergency arises. In that case, the protocol would mostly call on those machines that do not serve a critical function to their retail owners. In other circumstances, the protocol may call two of lets say, three machines, owned by a household or company that was made up of relatively younger people, rather than any in a household or facility of mostly the elderly, and yes the machines will return themselves in rotation, or when the duty service ends. Does that answer your question?", Ana nods.

Michael continues, "There are a few other areas that I would be reluctant to get into now, because of time, and well, the specifics are a little difficult. Suffice it to say that protocols at all the levels we have discussed, restrict and inhibit the ARCH 500 prototype series and 600 series machines from harming humans, and are tasked right out of the box with the help and happiness of the human population. There, hopefully that should answer that question that is always just in the background in most minds, but which, most people are too shy to ask, fearing being viewed as stupid and cliche. Now you can relax!"

Boirix interjects, "And just to clarify what Michael is talking about, we have to counter the perception from past media that robots will someday rise up and take over, and threaten the population as slave masters or other such nonsense. We have all heard it, and we are all adults here, but it would astound you how often that silly myth surfaces in discussions like this one."

The meeting is adjourned, and they all stand to stretch and mill about. Raising his voice, Boirix says, "Please, all, can I have your attention for just a minute. We will meet tomorrow at Grandpa Buckeye's at 1:00 pm to see more of our ARCH 540's demonstration. Ya'all have a good evening, and I will see you there tomorrow." Boirix, and James Bowen then exit the room.

Peter Zellhuber says, "Boirix asked me to take you all down and help you find your way round. If anybody has any questions, please ask me now. Does everyone know how to get to Grandpa Buckeye's right off Hines Boulevard..., does anyone?"

CHAPTER SEVEN

Bot V Bull

Grandpa Buckeye's

Michael is at the podium in a large hall filled with about 6 to 7 thousand people gathered to hear a continuation of the account. Since he started telling the story, he has become one of the most popular speakers, giving his version of history.

[MD Narrative] - It was a long time ago, but I remember the debut pretty vividly. I had been involved with the project for only a few years, and to me this was the finality of what had started, and my first real experience with seeing something of this importance come together. Looking back it was all very exciting.

§

Arriving at Grandpa Buckeye's, the parking lot is completely full, and parked vehicles have spilled into the extended dirt lots adjacent and up and down both sides of the street around the complex, which is not that unusual on a Saturday in Dallas Texas, except this is 12:30 pm in the afternoon, which is not the traditional time to start the festivities, but this is also not an ordinary day, because there is an extraordinary event about to take place. Driving through the parking lot, the magnitude of the event is evident because there are television news crews from at least 5 local news channels, and several others for Texas, and at least 2 from other states that have come to report the event planned. The news trucks are parked in many different locations in and around the complex, with several of them adjacent to the utility and staging area in the back of the complex, next to the rodeo pen and arena. There are balloons, banners, inflatable vinyl funhouse rides, ribbons and bows, bales of hay, cotton candy and a large assortment of creative deep fried delights, and clowns are present; all the things one would expect at a carnival event in the heart of dixie, except this was Texas, so everything was bigger than life.

This event was no exception either. Inside the main building, there are numerous venues for activities like mechanical bull riding competition, and instruction. Patrons also enjoy line dancing and live music events, complete with restaurant and full bar. Grandpa Buckeye's is a Texas style honky tonk, dance hall and bar in the heart of Dallas. It is a large complex with venues for rodeo performed every Friday, and Saturday evening.

Dalbots, already known around the local area for their excellent smart kiosk technology which is widely employed around the world, elects to introduce their flagship ARCH 600 series to the public at this location via a demonstration on a Saturday night in the middle of summer, 2028.

They are overwhelmed at the response they get at the event. They had talked a little about it on local cloud and broadcast radio, and even distributed a few fliers at the facility 2 weeks before, but this was overwhelming. In attendance with the Dalbots crew was a delegation of US government officials who had come out to tour the Dalbots facility, and observe the live demonstration. They also encounter numerous individuals, and executives from industry that had expressed an interest, or even had initial discussions with Dalbots about the equipment. The event was covered by the news from at least 12 local, state and out of state broadcasters, both cloud, and over the air.

Boirix, and Jim Bowen are there in typical Texan garb with ten gallon stetson hats, and rattle snake cowboy boots. Both are native Texan, born and bread, and both usually speak with a slight but discernible Texas drawl, but this is a special event and Boirix is around his people, and feeling his oats, so the Tex really comes out as he turns into a soggy Texan, minted in 1850, complete with all the usual bravado and swagger of the genuine article, as one might expect. Like with the rest of the regional cultures of America, a deeply struck icono-cleft, with split identity, having been brought up saturation bombarded by a media culture of sensationalized stereotyping and sham overwrought depictions, as so much of it has deluged the American landscape for so long, that even native Texans, having been threatened with the complete loss of cultural identity, are confused as to whether their speech, mannerisms, and customs, are genuine organic cultural artifacts, or just spurious manifestations of sham media conditioning. In any event, it feels real to them, and in this time of jubilance, and exultation with celebration, feeling the power of spotlight, as small as the event really was at that time, by success or adoration, it seems supremely necessary to display a healthy dose of ethnic-Texan pride. He is in front of a huge crowd, that includes government, and military brass, and representatives of Global Mega-corps, and from industry, and he is showing what he is about.

The venue is outside at the rodeo arena and the seats are nearly as full as they have ever been. The podium is arranged like an old time political event showcasing a presidential debate during the 19th century with large star-spangled ribbons draped in several loops across the 30 foot span in the VIP section of wood construction, which just received a fresh coat of white paint. The perimeter of the area outside of the wooden railing is standing room only with camera crews in various locations manning their telescoping camera platforms. They are there to report and record the event. This is the main event, the reason for all of the excitement, and the tenor of the crowd is powerful and unmistakable, and very evident by the pitch of chatter.

Gordie "Gordon" Hardesty, a tall thin gentleman in his mid 40s, who is the owner of Grandpa Buckeye's, is sitting in the VIP section next to Boirix and James Bowen. Behind them are several other Dalbots staff, including Greg Tishman, and Larry Bartelli, a few company staff, and several other business owners and executives from around the country that have come out by invitation from Dalbots for the event, like Frank Raney Jr. Gordie arises at the podium and starts to speak which begins edging the chatter down to a low roar.

He announces a few pieces of business and some of the coming events scheduled, then mentions a few things about local sponsorship, then proceeds to introduce Boirix, "From time to time, we have the rare opportunity to host something really special, and today is one of those rare occasions. The gentleman that you will meet tonight, and the company he put together a little over 12 years ago, Dalbots, which is already known to many of you, and around most of the world.

107

The Autonomous Economy I - The Strong Man

Today he has something special. How should I describe it...,
unbelievable..., fantastic..., extraordinary, historic, whatever
adjectives you want to use to describe it, it is here today at Grandpa
Buckeye's, for the first time ever showed in public, for ya'all's
enjoyment. And now I would like to introduce the man that created
it. So without any further A-doo, I introduce to you Boirix
Dalgleish..., come on up here Bo, everyone, lets give him a hand."

The entire crowd stands up and breaks into a loud and prolonged
cheer and clapping. Boirix stands up at the podium and pauses for
about 15 seconds waiting for the cheering crowd to quiet down.
Another 15 seconds go by, and the crowd is still cheering. Then
Boirix raises his hands and makes a downward waving gesture which
tells the crowd to quiet down. Boirix is quite taken by the response
and is not quite sure what to make of it. He had given a few
interviews to local news reporters in the few months before the event,
but was really not at all aware of the extent of the anticipation, or the
imagination this little bit of publicity had engendered in the local
population. It began to dawn on him that the company and the new
product were going to be getting a lot more recognition than he had
ever previously anticipated.

He begins to speak, "How ya'all doing today, boy, I'm not a public
speaker, but that was the most enthusiastic reception that I have ever
gotten, thank-you..., thank ya'all very much. As Gordon told you, I
am Boirix Dalgleish, Bo to my friends."

He takes a long pause for about 20 seconds to form his thoughts,
then continues, "What you are going to see today is the result of
about 8 years of intense research, development, and sweat by my staff,
they are who you should cheer for, several of them are sitting with me,
and right behind me up here, come on and give them a hand, they
deserve it more-so. Especially our chief scientist and tech-magician
Doctor James Bowen, Jim stand up, let the crowd see who you are.
Jim is a founding member of the company, and he has been steady as
a rock in bringing the project to its present state, and we would not be
where we are without the dedication and perseverance of Jim Bowen.
Join me in giving a hand for Jim, and the Dalbots staff."

He begins clapping for the staff and the crowd joins in and stands
to their feet for about 15 to 20 seconds of enthusiastic clapping and
cheering. Boirix then gestures for them to quiet down and continues,

"As Gordon said, you here today will be among the first people to see the new prototype that we have brought out here for your inspection. I gather many of you are aware that this is brand new technology and has the potential for revolutionizing our world. As we commenced on this endeavor over the last several years, It began to dawn on us how what we were creating could have a very positive multi-dimensional impact, and enhance the lives of ordinary people, and that pleases us immensely.   Now I would like to introduce to you the Dalbots ARCHIE 540 prototype. I think he is going to amaze you. Oh wait one minute, I forgot to mention that after we are done here, which they tell me is about 2 hours from now, we will move inside to complete the demonstration."

Boirix is distracted and talking to Gordie sitting next to him. He is busy asking where inside to go, to relay that information to the crowd. Boirix then continues, "We will be moving inside into the 'Air-Conditioned Bull-Pen' to complete the demonstration, now lets get the ARCH 540 out here."

The crowd once again erupts into standing cheer as the strange looking machine walks out into the middle of the arena and begins to wave to the crowd.   For about 2 minutes, the ARCH 540 walks around just inside the perimeter of the arena in a perfect circle, while waving at the cheering crowd. The machine completes the tour and walks back into the gate from where he had come out.   Within 30 seconds, a gate flies open and a large bull erupts from the staging area and begins bucking about, as a slight but muscular young man on his back, hangs on with one hand held in the air so as not to be disqualified.

He manages to hold his grasp for 8.5 seconds which is displayed on the lighted score board rapidly ticking by in 10^{ths} of a second. It also shows his name, Billy Crager, and the name of the bull he is riding, Mean-Streak.   In the middle of the display is the scoring section giving a final score from 8 figures total, for both the rider and the bull, entered by 2 judges located in an elevated loft at a far end of the arena, along with the booth for the announcers.

The rider has a number pinned to his back, Billy Crager, Number 18, is a local champ and is homed from Grandpa Buckeye's.   He competes statewide and is a rising star in the southwestern regional division, and has competed internationally in 23 events.

The Autonomous Economy I - The Strong Man

The loud-speakers boom as the announcer details the particulars, "Billy Crager on Mean-Streak, a crowd pleasing Grandpa Buckeye's favorite, oh, oh and he is in the 8s', there he goes, oh, endo. Billy is an up and comer, four years with ipra, or IPBRA of which Grandpa Buckeye's hosts many events each year. Billy is twenty and has an impressive average for a man as young as he is. He has competed internationally for the last two years, and Billy's score, eighty nine and a half; nice showing Billy."

After the bull dumps him, as he lands, he bounces along for a few yards as if his body was made entirely of rubber, and as usual the rodeo clowns and safety staff are there to distract, and release the flank-strap from the still flailing bull as Billy makes his escape. Billy then stands erect and brushes himself off while exiting the area. The next event is number 15, J Thomas Jackson on Lightning-Bug.

The announcer belts out the action, "One of his most favored bulls to compete with, Tom and Lightning-Bug are old buddies."

8.8 seconds and Tom lets go, rather than being tossed. He rolls off Lightning-Bug's backside as if it was planned and is only slightly air launched about 8 feet before landing with a loud but softened thud from hitting the mulchy ground lining the Arena. He rolls again to avoid injury.

Peter Zellhuber, Michael Dalgleish, another engineer, and 2 techs are in a covered part of the staging area, at one far end of the arena next to the main set of bleachers where the bulls and riders are let out into the arena. They are discussing the strategy while the ARCH 540 sits on a bale of hay near a table holding equipment attached to him. Both of them are nervous that something may go wrong with the live demonstration in front of an audience, while both Boirix and Jim Bowen are counting on a perfect demonstration. Although, they have been working diligently for months to work out any bugs in the system, and have been practicing what the machine is supposed to perform, the odds that something will go wrong are better than even during a demonstration, and quite often it does.

Peter is making some last minute measurements and is somewhat agitated as he speaks, "Michael have you run the last set of diagnostics yet?"

Michael responds, "Hold on to your hat, I'm just about done. I told you I ran it earlier and it was fine, no problems."

110

Peter talking to himself, "I cant believe things are going right for us, things never go right, there is something wrong, things always go wrong at the last minute when giving a demonstration, its the immutable law of demonstration. What is going to go wrong?"

Michael asks, "What are you babbling about?   Did you say something?"

Peter responds, "Nothing man, I'm just mumbling to myself.   Ok, our boy is up next. Lets get him up there."

Peter monitors the remote console so both of them can watch what the ARCH 540 sees and does as the machine walks confidently into the pen and mounts the bull. The pen doors fly open as bot and bull lunge forward.

Back outside in the arena the announcer is finishing the last rider's performance as he belts out, "And he scores a 91, an impressive ride indeed by this young man. Up next, uh-oh, and here we go, this is Archie Dalbot on Banshee. This is truly a first for us here tonight folks. Banshee is admittedly hormone enhanced, and known to have maimed and injured so many riders that he was considered to be, 11, 13, 14, 15, is he even..., holding..., 18, 19, ladies and gentlemen, Dalbots' Archie is hanging on like nothing we have ever witnessed, thcy have released Banshee's strap..., and the Dalbots entry..., 20 seconds on the ride, as graceful a dismount as has ever been witnessed in Professional Bull Riding history. Ladies and gentlemen, you have just witnessed a historical event.   Looks like he could have ridden Banshee forever. We have just witnessed history being made, and the judges have given him a 97.5, unbelievable, unbelievable!"

97.5 was the highest score ever officially awarded at an event in the history of the sport, or a least that was the rumor. After 20 seconds into the ride, the safety staff decided that the ARCH 540 was not likely to get bucked off, and the bull, is starting to tire out, so they decide to release the strap. At about the same time, the ARCH 540 let go, and as the hind legs of the bull bounced up and to the side, the ARCH, having measured the frequency and magnitude of the buck, computed a probability vector and trajectory of the next launch for the hind-end of the animal, then maneuvered an even higher stance using one leg, and pushed himself off the bull's back into the air. While in the air, the ARCH did a tuck and double roll, perfectly landing with a sprint about 15 feet away from where the bull is still bucking, and where the crew is still attempting to relieve him. The crowd goes wild with applause and whistling for the next 1 and a half minutes without a break.

The bull riders who went before the Dalbots entry were looking rather sheepish, and insulted with long hangdog faces. They earlier, having thought that their particular performances were very good, were feeling very confident. Billy Crager especially thought he would place higher on the purse.

He, and three other entrants later give an interview to a few of the reporters in which they complain that, "The rules for Robots are not clear or even legal!", therefore they maintain, "The judges should eliminate the score of the Archie Robot."

An attractive young brunette female reporter is holding a microphone and speaking while interviewing some of them, "This just seems like sour grapes from these young competitors having to compete with a machine for the money of the purse, even though we have been assured by association officials, there are absolutely no prohibitions or qualifying criteria for non-human contestants. This is Wendy Taylor-Decker with Cloud 9 news, reporting live from the Grandpa Buckeye's arena in Dallas."

Again the gate flies open where beast and man emerge as the announcer speaks, "Up now a first time for this young man. Ngambo Mbose from Nigeria on Chug-a-lug. Oh and he is disqualified..., and he's been flung. Cant touch your face Ngambo. That is a lot of wild bull for a newcomer. No score."

Mech-Beast

The next event that was scheduled for the Dalbots crew was inside a large building in the complex, so the announcer began to let the attendees know where to go, "Folks, if you have come for the Dalbots event, you need to move into the AC Bull-Pen in the main complex. The event is scheduled to get under way there in about the next thirty minutes, so if you are going to go, you need to start making your way over there now. Repeating, if you want to attend the second Dalbots event, then start over to the AC Bull-Pen now to get seated. For those of you staying with us, our next entrant is a young man who hails from colorado."

Nearly 90% of the entire arena, which holds 2500 people, and was nearly full at the time, left to go inside for the event. Inside the AC Bull-pen, the atmosphere is very electrified and excited with wall to wall standing room only, with some left outside in the overflow area, and some going into the bar and restaurant to watch the event on a large television monitor. The building is large and made of metal with a 30 foot ceiling, the whole room is about 50 by 75 feet with bleachers set along 3 sides. It occupies about one third of the building.

Out through the hallway leading into the room, is the indoor arena for cattle and livestock auctions, calf roping, and barrel racing along with other events that are usually staged inside. In the middle area of the building is a ring with a mechanical bull in it, however this is not the usual mech-bull, because the Dalbots crew had recently taken it back to Dalbots to modify it for the event, then brought it back. It had gone missing from the ring for a few weeks, taken back to Dalbots where modifications to the original chassis were made by the engineering staff. They beefed up the entire system by adding a new speed level, higher than the maximum from the factory, which makes it much more difficult to hang on for the requisite 8 seconds in order to qualify for the round.

As one might expect, there were a host of competitors itching to prove their stuff with an abundance of the stupidity and bravado that comes with youth, all lining up to compete against the Dalbots entry. Many of them had various colorful observations, brilliant insights, quips, and quotes for the extensive media presence, which was crawling all over the place, attempting to interview anyone that would care to give a comment.

"I can beat that robot thing because I don't think it has what it takes in the long run to compete!",

"I'll make scrap metal out of that thing!",

"There is always a flaw in machines, they're not as smart as humans!", as if brains were among the primary requisites for competing.

"Does it have a battery or what?",

"I know some people that have one of those, and they let the kids play with it all the time, I wasn't expecting to meet up with a robot here, or compete with it.",

The Autonomous Economy I - The Strong Man

"It was an awesome machine, and a tough competitor.",

"Got my ass handed me!",

"My wife's washer and dryer can do more than the Archie Robot."

It was not long before the 1st of 9 of these brave boys and 1 girl stepped up to compete and ride the modified mech-beast for time and style against what they all knew was the odds-on favorite entry from Dalbots, but the stakes were high enough that, even though there was some fear and hesitation by some, after witnessing the afternoons developments, greed, ambition, and the desperation for money soon overcame their apprehension and the game was on.

The announcer, who also works as the auctioneer for the complex explains the rules and regulations, "Welcome contestants, welcome all, welcome! This is a special event tonight. We have about 24 entrants competing for a sizable purse, and we have a very special player here, the Archie Dalbots entry that some, a few, most, however many of you saw this afternoon in the arena. Just a recap of the rules for the association, to qualify, you must be 15, have registered, including any waivers, and have paid the entry. Repeating, you can not compete unless you have registered and paid the entry. Contestants, meet Grinder. Contestants, you must also fall into one of the following classes: Novice Beginner; Mid Novice; High Novice; Low Amateur; Mid Amateur; Semi Pro; Professional; National Champion Class; International Champion; Supreme champion; and one new class we are adding that the association has not yet sanctioned, and so it will not count or record until the association adopts it, nevertheless we are adding it to the roster tonight. The new class is called the Ultra Supreme Champion class; ultra supreme champion class!"

The announcer continues, "Now I need to advise you that Grinder, the bull here tonight, and the ring have not been inspected and certified by the association due to time constraints, since it has been modified, but we believe that both are otherwise regulation compliant, and will be found to be so by the association. Whats that?"

The announcer is interrupted by a gentlemen speaking to him. He covers the microphone with his hand for a few seconds, then goes back to speaking, "I'm being told that there is an association representative here already, that has allowed the contest to proceed with the modified equipment. Both the bull and the ring have been modified to reflect the new ultra supreme champion class which we are adding tonight.   Now, currently none of the entrants qualify at that level save the one special entry, and we all know who that is.   Let the 1st round begin!"

Contestants 1 through 4 were dispatched somewhat quickly with only one of them receiving a score, and that was a 68.   Billy Crager was the fifth contestant to step up to ride the beast.   Billy having competed since the age of 17 had some talent and was favored by the crowd that chanted his name as he was up and riding the mechanical menace.   He was expected to score high in the contest, and with consideration of the level of competition, and the reality of things being what they were since the afternoon, he did not disappoint, at least in the early heat.

Billy registered, and qualified in National Champion class and so that level of difficulty was set at the controls for the bull, but in the final heat, he decided to raise the difficulty one level beyond his qualified class, which in normal circumstance would have been denied by the rules, but due to the media attention, and the special entrant, the rules were waived after both he, and his manager signed a special waiver.   This was a particularly risky move on his part because the money can only be awarded based on the level that he would qualify for, and not the level he competed at.   This move was based on his desire for notoriety in the sport, and the media was there covering the event. A few days later, after the excitement died down, while lying in the hospital with both legs in suspension, and after 6 hours of surgery, having received 3 metal rods and 13 screws, he realized that he had made a miscalculation and it cost him a considerable amount. Contrary to what Billy Crager thought, the media was only interested in covering the Dalbots events, and in their coverage, did not mention his performance, nor his blunder, nor that of any of the other contestants, other than to play a few very short clips of some remarks from a few of them.

Ironically, to his chagrin, a few days later as a follow up story to the Dalbots exposé, a reporter showed up in Billy's hospital room and was granted an interview with him in which he did get some notoriety, both about his stupidity and about his ability in the sport.

The reporter asks, "Billy what did you think of the Robot?"

Billy's response, "I thought it was the most incredible thing I have ever witnessed, I cant wait to get back on-back again against another one!"

The reporter closes, "Well there you have it, the exuberance of youth. Back to you Paul!"

In the heat, Billy was thrown clear of the safety barrier that consisted of very thick padding that lined the retaining pony wall and the floor. He landed mid way up the bleachers, as the crowd scattered to get out of the way of the human cannon ball hurtling at them from a considerable distance. He then bounced and rolled off the edge and belly flopped onto the concrete floor below. He lay there surrounded by concerned onlookers, and the onsite medical staff attempting to asses his injury and render aid until the ambulance arrived. Thankfully there was a hospital only 4 miles from the complex, which is always at the ready and knew exactly where to send the ambulance when the calls come.

Also during the event, the ARCH 540 rode the machine at the highest level, the new so called 'ultra supreme champion class', which was created by the Dalbots team modifying the Grandpa Buckeye's mechanical bull. The ARCH 540 rode it gracefully, making it look easy, and the crowd cheered wildly for the 28 seconds the heat lasted in which the ARCH 540 held on, and did not let his hand touch any part of his body or that of the bull. After the ride the announcer commented, "I wonder where he learned to do that? Who taught this robot to ride a bull, where is this school? Folks, I think we now know where the Grandpa Buckeye's mechanical bull has been for the past several weeks, and who's been riding it, cause it sure wasn't here."

Huevos Toritos

The Dalbots group that were invited, or that otherwise attended the unveiling stayed after to enjoy some hospitality in the Grandpa Buckeye's complex restaurant, complements of Dalbots. In Texas, many of the people have very hearty and earthy tastes. On special occasions such as Saturday night at the dance, quite often many are in the mood for a treat, so there is a very popular earthy appetizer, and they bellied up to several large table size platters of what are euphemistically referred to as deep fried Texas Orb Steaks, or Dangle Steak, or even Golden T-Nuggets, huevos toritos, Texas Bull Giblets, etc., which in other parts of the country are variously referred to as

Rocky Mountain Oysters, Tennessee Turtle Eggs, and so on. These were sliced into round flat portions, floured and deep fried whereupon they would warp and twist and could easily be mistaken for chicken, gator tail, fish nuggets, or even rattle snake chunks with your choice of dipping sauce, which ironically was also on the menu and the 2nd most popular runner up to these gems.

Peter Zellhuber was adamant, making sure that everyone uninitiated at the table was sure to try one. Bill Dubrowski and Grady Hunt, who's regular pampered diet, served daily in Washington is prepared by imperious culinary snobs with grandiose pretenses of stature and what ranges from overcooked to virtually raw foods, served on white tablecloths with expensive wine; quite formal and stiff, who although aware of what these delicacies were, tried them without hesitation, not wanting to look weak or feeble but strong and virile. Grady Hunt felt quite comfortable downing 4 to 5, especially the Tabasco fire-glazed version called Buffalo Dangle Steaks. Ana Isidora, who was easily satisfied regularly with a tuna salad sandwich, was quite taken with them, especially the naked ones that she would dip in the marinara until she was told what they were. At first, unbelieving, but then it began to dawn on her that she had heard of these being eaten, and so she decided she had had enough with just the 2 pieces she ate, although she was quite silent and little pale for the rest of the evening.

General Cranket was quite comfortable, and being a Texan, he had eaten them since childhood in every conceivable way they could be cooked, from slow roasted, smoked, stewed, slow braised, and deep fried, even boiled as unappetizing as that sounds, and believe it or not, in Texas they even barbecue them. However they come, the General has put them down. The platter was handed to Frank Raney Jr. who turned them down flat complaining that he never eats any "Strange Meat."

The expo event at Grandpa Buckeye's was a monumental success, and helped Dalbots launch their latest flagship offering far and wide.

Apprenticing Machines

For many weeks and months after the company's unveiling at Grandpa Buckeye's, the media had become obsessed with the company and the ARCH 600 series machines. Several of them were a constant fixture at the company headquarters, requesting interviews, more demonstrations, and more information about the scope of the technology, which Boirix and the staff at Dalbots obliged gladly as often and as thoroughly as they were able. The extent of media coverage was debated by many in the company including Michael Dalgleish and Peter Zellhuber who were very cautious and skeptical about being too forthcoming and open with the media about the technology. They feared that while the old presumed axiom of 'any publicity, even bad publicity, is good publicity', is accepted, in this situation, it is probably folly and there is a price to be paid at some point for the so called 'free publicity.'

Sadly this fear was realized multi-dimensionally and in numerous instances and circumstances. There was some immediate fallout as competitors and those motivated by the theft of the technology, were soon attempting to blend into the frenzied media hordes in order to get some inside line on the companies secrets. Some approached the company ostensibly as journalists in every conceivable field, but were in actuality investigative agents of competitors and foreign governments alike. Just as any time in history, when bold and life changing initiatives are undertaken by visionaries, of which Boirix Dalgleish certainly fit, there is always considerable detractors and critics with nothing constructive or complementary to offer, who will instead attempt to tear down and denigrate the effort. There was also considerable criticism, mockery and outright hostility from several quarters reflecting a myriad of motivations and sentiment.

Week after week, the local cloud-caster Cloud 9 news was an almost constant fixture at Dalbots, and their field corespondent Wendy Taylor-Decker was spearheading the media penetration by Cloud 9, "This is Wendy Taylor-Decker coming to you live from Dalbots headquarters in Dallas. We are talking to the companies Chief Scientist Doctor James Bowen who is explaining the basics of what the future may be for many in the manufacturing and services sectors of our economy. Doctor Bowen, can you tell us what is happening in the background?"

Inside the training & development lab in the R&D department of Dalbots, the ARCH6 machine is attached to a tether and is duplicating the movements of a training operator who is wearing a special bodysuit trademarked as Jumpsuit™ which senses the position of the wearer in 3 dimensions and in real-time, including the position of the head, the fingers, hands, feet, mouth, eyes, and virtually every muscle in the body of the wearer. It is called a WYDIWID generator, for 'What You Do Is What It Does', and is also sometimes called a wydisuit. The movements and positions are fed into a data acquisition and vector computer produced by PedF SA or *Penser et de Faire*.

The hardware-software system for training generates program sequence steps from the suit. The entire system is called JumpStart™ and is used to train the machine to move in a coordinated manner, memorize the movements, and use the programmed moves as an adaptable interpreted template, for which the resulting output sequence and operators are compiled into one or more of the X-Packs™. So when the trainer is performing, it does what he does. It operates in both a simple fixed program mode or recipe mode, whereby the learned movements are repeated exactly and flawlessly; or in an autonomous mode, whereby the machine will attempt to execute the movement sequence, but may modify the movements by taking into account dynamic contingent real-world conditions, such as reacting to other things that move, or in a setting in which the operational objects may not be located relative to where they are during the programming-training procedure.

As an example, a machine may be trained to pick-up an object with a right hand, then pass it to the left, and then set it down to it's left, as part of a more elaborate sequence of a more complex task. But, attempting to perform that higher level complex task in the real world may be hindered if the objects of the operation are not located exactly in the same location relative to where they are placed during the training. The way this problem is overcome, is to use the ability of the machine to adapt and make decisions about contingent conditions that can not be anticipated during a training sequence, which allows it to modify the sequence as conditions warrant, but still accomplish the task timely and accurately. In this way, the machine becomes more autonomous and universally adaptable to the requirements. Instead of the user having to train a blank machine for each routine task, the machine comes with the knowledge and skill 'out of the box', for certain simple to more complex tasks, and with the ability to adapt to non-standard conditions for which it has not been specifically introduced.

There is a general word for this phenomenon which is 'determinism', and is regarded in high level courses in most schools of science and engineering. In the context of robotics and synthesized intelligence, determinism has to do with how a machine will act given a task with a certain environment and set of circumstances. A highly deterministic system will act less autonomously with less ability to compensate for circumstances which it has not been specifically trained, or has not overcome before, while systems like the Dalbots machines are highly non-deterministic, meaning they are extremely adaptable to unknown circumstances, which ideally suits them to replace many human laborers performing low to medium complexity work.

The earliest units employed were able to perform a given task similar to the way humans do things when performing repetitive tasks. In practice, with humans, it is necessary to limit the number of such unanticipated contingencies, by limiting the scope of the work, due to the fact that humans must then exercise judgment and make decisions which ultimately increases errors. The problem is much more easily overcome with machines because they can be given the ability to learn and adapt to simple work at a rate greater than that of humans, with greater accuracy and speed in the operation. Also, there is no limit to the corrective-adaptive modifications that can be performed with machines, something that is certainly very limited with humans. Through the X-Pack™ facility, this knowledge can then be cumulatively compiled, copied and transferred to other machines, whereas that can not be done with humans in any reasonably comparable way, but only a very marginal adaptation through much cost and aggravation.

Reporters and news organizations of all sizes and shapes reported on the progress of the company in numerous situations and on various occasions. They reported on the company while developing the job training protocols.

The media vignette and many others were cloud-cast repeatedly, and watched by millions in most states and foreign countries over the succeeding months and years, which wetted the appetite for other media channels to also cover the company and the ARCH. Although there were other entrants in the industry, and a few attempts at copying the technology, the Dalbots ARCH 600 offering had successfully leaped way out ahead in the forefront, and captured the hearts and imaginations of an admiring public.

Jim Bowen explains, "This is our training lab in the R&D facility, and we are going to observe some machine training that is being conducted by one of our research staff."

The reporter asks, "What's that he's wearing?"

Dr. Bowen answers, "He's wearing a training suit as you can see. He's a training operator and the suit is a PEDF JumpSuit. OK now watch. While he's holding a paint brush, which he sets down on a table, then picks up the can of paint, notice the machine does the same. Now as he opens a can of paint and begins to stir the contents; now he is poring it into a tray."

The machine clearly and cleanly follows the exact movements manipulating the same implements. Jim continues, "The machine is being trained by the operator. As the operator moves, the JumpSuit generates the motion signals that are fed into the machine, commanding it how to move. The machine then follows the movements of the operator. The machine remembers the moves that it has been taught, so that it can repeat the sequence at any time."

Observing the report via the monitor of the news crew or while watching live, the reporter is speaking into the microphone and in front of the camera. She then wraps up with remarks, "From the heart of Dalbots in Dallas, there you have it, right from the company's chief scientist and developer, I'm Wendy Taylor-Decker reporting for Cloud 9 News."

Successive media stories followed the progress of the ARCH 600 machines as they were being delivered and deployed by numerous commercial customers. By 2031, just 2 years after the unveiling, there had been near 25,000 machines delivered and employed in various disciplines mostly in the hospitality, leisure, manufacturing, agricultural, retail, construction, labor, and government sectors, then a decade later, they started showing up in the higher disciplines.

CHAPTER EIGHT
Deconstructing Man

[MD Narrative]: When endeavoring to create something new, it is helpful to have an analog from nature to serve as a model or guide. A model is a kind of roadmap for those that have never followed the particular course of discovery, so a model can serve to give many ideas, lead to many discoveries, and flag dead-ends not anticipated. I know of which I speak. I know of feeling the way through the dark. I have been there at the point of discovery; of creation, when something never seen before was born, in which I had a hand. In so endeavoring our creation, it helped that there was an analog from nature to use as a template, a kind of prototype. A few examples are: the bird as a model for an airplane; fish modeling boats, submarines, or swimming apparatus; walking, slithering, crawling creatures like snakes, crabs, and spiders for their mechanical counterparts, and so on. So it follows that a composite of the modern human being was used as the analog model for the humanoid robot.

Extracted from the human analog, the machine may resemble the human in many aspects like body type, number and location of limbs, head, torso, and so forth, as well as in many aspects of cognitive and thought functions. However, in many critical respects, the resulting creation, falls way short of the model.

Is it still a work in progress? In many ways it is. There are aspects of the model for which, after numerous approaches, and considerable effort by many endeavoring to understand and complete the work, the best attempts to emulate are poor and very superficial, and in my opinion, these aspects are critical, but their creation is not within the power of men to perfect. These aspects are outside of the physical world in which we have access, and working within the required realm to perfect them, penetrates esoteric mysticism. For our purposes, it is constructive to make comparisons for discussions and understanding of the relationship man has with the machine model, his environment, the world, and the vast infinite universe, thereby dispelling darkness from the path of discovery.

Man: Mind, Body & Soul

[MD Narrative]: When it became clear there were critical attributes of the machines we were creating, the so-called perfect model, and that what we could produce would fall way short in many respects, there was an effort, or better, there was an imperative, a number of years ago, initially in the early-years and then again later, to engineer the physical model, but also to emulate aspects of the human spirit and psyche. So an initiative was started, by consulting the then known science of the psyche, and spiritual professionals, to better understand those aspects. First was conception, then data was gathered and incorporated into the design. The objective was to arrive at some understanding of how a man was constructed, but not get bogged down in sectarian squabbles or fine points of theology. In addition, what was then to be the result was to be drawn on such lines as to not broach religious dogma of any kind, but to present a thorough treatise, in mostly scientific or secular terms. That objective came to be known as the 'Deconstruction of Man', or again, 'Deconstructing Man.' My father spearheaded the effort after consulting his own minister, whom he asked to join.

The following was put forth by a working group that issued a report. There were essentially 2 groups; The Psychologists, Engineers, and Behavioral Scientists formed one group, while the Theologians, Shamans, and Gurus formed the second; collectively they constituted what was called the working group. Although they worked separately on their own, and communicated as necessary, at some point they combined their findings together into a report, a theory, which they were able to present, constituting what they could agree on from the standpoint of understanding what man: mind, and body, are. That was the simple part. The soul of man, that was the challenge. Did it exist; was it mythology; how did it function; what part of it was a concern, or necessary to understand for our purpose; and can we emulate the important components, or relevant attributes, to bridge the gap in the design? These questions were asked of that group, and they presented parts of their answers separately from the group as a whole to avoid intractable differences. A separate initiative called the Spirit Inculcation of Sentience Initiative or (SISI) was covertly started after the findings by Boirix, his minister, and a small select group within the company.

Although, thought to be critical, the initiative was abandoned after it became clear what the scope of research entailed. Then recently, it was re-initiated after it became clear and we witnessed the full scope of the earlier failure. Again, the spiritual profession was consulted and we were in luck because they were one of the only professions still in practice at the time, but that is another story for another time.

§

Having conducted much of the original research into making an artificial man, and now having progressed considerably in the undertaking, the research group at Dalbots are still involved in various aspects of undefined portions of a complete engineering specification, the completion of which will define the scope, and reveal boundaries on the project. Boirix has decided he needs to have a conversation with his presbyterian minister, in which he hopes to explain the nature of the questions, solicit answers, and get help in an initiative that will further define what they have to do to reasonably accomplish the objectives.

Boirix and Celia Dalgleish are friends with the Reverend Bertrand Oates and his family, having known the reverend for many years. Both of them regularly attend the church he pastors. Boirix meets with the Reverend at his office in the church and explains what he is up to, "Bert, I was hoping to get your help and input on an important aspect of a project were we seem to have run into some difficulty. Do you remember the project my company is undertaking?", he replies, "Robo-man, how could I forget that, I found it extremely fascinating, and was wanting to ask you how things were progressing, and to be perfectly honest, I was wondering if you might not be asking me for some guidance sooner or later. I guess my hunch was correct."

Boirix continues, "As you may have guessed, there are some areas we are not really clear on, where I thought maybe you could shed some light. We have done research to flush out the biggest part of the model, and have a pretty good idea for modeling a person. We have looked at the physiological, and intellectual aspects, but I have come to some interesting questions about the interaction…, our basic problem is one of defining the extent to which we have to restrict autonomy and simulate volition, and came up against what we have defined as moral judgment, and accountability, we figure that means some spiritual aspects, maybe soul, I'm not sure of the extent of what we do not know, and need to consider. I brought this whole topic up to my team, and one of my chief architects, Peter, is not really on board. He thinks the idea of a soul, or anything outside of what can occur through evolution is, as he puts it, 'garbage', and has not been hesitant to say so. He thinks we just open all the stops and let mother nature take its course. I am not confident in that approach, we need some clarification."

He continues, "We have followed the design model to accommodate the labor enhancement of people, and while we have modeled aspects of the impact on society at large, well, there are some obvious protocols that must be handled correctly or we have a problem. We need to know, what is important, and what can be modeled. I thought you might be able to pull something together, an interfaith group of sorts, who can add some insights, and issue a report?"

Reverend Oates asks, "Are you playing God Bo? A man is a complex organism..., or maybe mechanism to the extent that applies here..., much more-so than a machine, and not something that crawled up out of the mud either, as they teach the kids in school. No I'm afraid, our view of this question is the unvarnished truth, although not all that popular today, but I suspect you already know that, and that is why you are here..., and I suspect that in the long run, you will find out why we are so adamant about that view! Look, we can probably get some answers for you, but it may take a few weeks. Interfaith huh..., I can pull together a few clergy from a few of the other reasonable denominations, and I am confident we can address your questions, and package something!"

Boirix replies, "Reverend, I am sorry, but by interfaith I mean it can not be confined to just presbyterian, nor even just the christian church. This is government sponsored, so it has to involve others, and must be dressed in non-sectarian and secular language, although, just between you and me, this is off the books for the time being, nonetheless, if that ever changes, I have to cover my considerably exposed backside on this one.", he replies, "Others huh, you mean heathens, and damnable cults?"

Boirix replies, "Yea, I'm afraid so, but I thought you were pretty fair so I figured you'd be able to handle it."

The reverend replies, "Well yes, we are fair, and I'm sure we can accommodate you somehow, although, I can not guarantee we are going to agree on things, nor be able to find a common ground. The more outside the pale…, the weirder we have to get, the less we are gonna be able to find a common denominator, nor a coherent or complete thought. I just want you to know that, but you know we can always get you to the truth, even while those lost for eternity go off the range. Why don't you work up a list of the type of questions your people need to have answered, and let me work on putting together a proposal."

Boirix replies, "I appreciate that Reverend!"

The group is pulled together by the reverend, and then is gathered for the first time to meet with Boirix and a few from his team, sans Peter who refused to set foot in a church. They have come to consider the scope of the proposal. A group of lay ministers, pastors, priests, monks, and assorted spiritual professionals walk into the church and toward the back, into a large conference room where they are greeted by Reverend Oates. They take their seats, ready for the conference. Reverend Oates walks to the front and greets the group, "Welcome all, and thank-you for coming. I have representatives from Dalbots that would like to address you and ask for your help. Some of you know Boirix, and know about his company, and some of you will meet and hear from him for the first time this afternoon. Gentlemen please welcome Boirix Dalgleish."

Boirix stands up and walks to the front and begins to speak, "Gentlemen welcome, and thank you for giving up your day. You may find there are many reasons for participating in the project with us, and some of you may object to what we are doing, and we respect that, but I am here to ask for your help. I only ask that you listen and withhold judgement until I have made my case. My company is involved in some rather unique research, and some of it breaches the realm of the spiritual and that is why I am here, and why you have been invited by Reverend Oates."

He continues to speak and lay out the questions at issue, then they exchange questions and Dalbots gives the group whatever answers come easily. They break and plan to meet again in the next few months. A consensus of theory was formulated, then compiled and presented. There was a meeting at Dalbots to discuss the report.

All of the Dalbots research and design team attended including Peter who initially sat with his arms crossed, with a very skeptical look on his face like, "I'll listen but I don't expect to hear anything except whatever confirms my belief that this is all a load of crap."

Peter was harboring much fear at what he might hear and have to deal with. Afterwards he commented that it was, "less crap than I expected," and he basically agrees with the parts of the report which were presented in mostly secular terms. While speaking, the spokesman for the consulting group presents using an electronic presentation board while others pass out a copy of the report. The report is marked 'Confidential.'

"We have created a report on the issues you requested, and have titled it, 'The Deconstruction of man.' On basic agreement, we have laid out the following premise and description: Man is composed of the constituent components of body, mind, and soul. The mind of man, exists in two distinct worlds partly through a cosmic singularity of sorts, a single dimensionless point, an infinitesimal pinhole that penetrates the barrier between the infinite cosmos, the mind of God, and the finite bounds of physical reality. It is part of the connection man has with an infinite God, the other part being the heart of man, wherein lays the volition or will, connected in similar fashion, and both of these allow for limited transcendence of some of the attributes of God into man. Man then bears in his body, mind, and soul, some of the attributes of God."

The spokesman continues, "The body of man is for all intents and purposes, a biological machine and behaves according to the laws of physical reality. The brain is the abode and seat of the mind, and acts as the biological '*hardware*' that filters, parses and sorts sensory input, and performs a similar function for outputs. The brain also adds physical memory necessary for cognizance and thought operations. The human brain can compute or calculate in the sense of weighing information by making comparisons with information stored in memory, and outputting decision or judgement. It performs many other similar computational feats in the process of control and regulation of the body, and giving the individual the ability to decide matters based on critical information. The brain also initiates, regulates, and controls many other vital and necessary functions, some involuntary such as heartbeat."

Continuing, "The body transports the brain. The body supports functions necessary to keep the brain alive, such as blood flow that carries oxygen and food nutrients to nourish the cells. It is thought

128

that cognizance, defined here as awareness of ones self, and surroundings, and the seat of volition, the will, continues to exist after separation from the body. When the body dies, the mind looses contact with the physical world, so at this point, it exists strictly within the cosmic side of the interface. In other words, it no longer is split between the spiritual and physical worlds, but now has been severed and exists completely in the spiritual realm."

The presentation continued to completion, and then there was some questions from the Dalbots group. Boirix asks, "The most obvious question, which I am not sure you addressed directly is what is a soul, can you give me a definition in more or less physical or secular terms, I mean, how is it constructed? How is it manifest? Do you have an opinion on whether it is necessary that we model something like it for our purposes, and how would you approach a model?"

Peter interjects, "Bo, you are way out there with this thing. We are not creating a race of fully sentient moral beings. As the reverend asked you, we are not playing God. These are simple machines, they are machines, regardless of whether men have souls, the machines we build are like animals that do not have souls. Why are you obsessed with that question?"

Boirix defends himself, "I have reasons that I have not fully formulated in my own mind, nor expressed to anyone, and I am concerned at some point, this is going to matter much more than it appears at present. I'll just have to leave it at that."

Reverend Bertrand Oates then asks, "Gentlemen, there are some things with which we, nor anyone else are going to be able to give you satisfactory answers for what you may think you need to know. Boirix, I am also a little curious, and am not defending Peter, nor his idea that you can simply let them run down whatever course nature will take them; I am not comfortable with that at all, but you need to be much more specific with your concern about the need to create a software soul. It is not clear, and I think that is some of why Peter is confused."

Boirix replies, "Thank you reverend, I think I would like to discuss this more in private with you until I am able to really sort this thing out in my own mind; lets just the two of us meet in my office when we conclude, and I can get a little more specific. Are there any other issues for the group before we wrap up? I would ask all of you in the consulting group to plan to meet at least once more, as we formulate ideas and questions. I think we on the design team need to thoroughly read through and digest the excellent report the group has prepared for us, thank you all."

Soul Machine

[MD Narrative]: All advances of mankind came about because of the unquestionably spiritual attributes and abilities that were imbued into man by his creator. Man was not able to likewise pass these attributes on to the machines, and these attributes are strictly in the spiritual realm, so can never be possessed by a machine. A machine will always be just that, no matter how complex it's algorithms, it's computational density, or how vast the network it traverses. The nature of a machine has been known since the beginning of recorded history and does not change. It can never have a will, or experience self realization, and it is not capable of moral decisions or actions. It will never experience curiosity or proclaim to itself…, 'I wonder!'

A machine is not, nor will it ever be a moral creature, and can never attain to the status of free-agent, and thus can not sin against God or Man, and so can not experience shame or remorse, nor repent. It can not experience inspiration and can not perform a single act of creation. Anything that may be mistaken for moral behavior or inductive thought or creativity by a machine is simply a pale facsimile of the nature of man impressed into it, and the artifice of man that emerges from it. Just as man can never attain equality with, nor embody all the attributes of God, so machine can never be equal to man. That which is created can never be equal to, or greater than that which created it.

As this axiom began to dawn on researchers, they started to realize the limitation, and conclude that your toaster would never be able to love or appreciate you because a toaster is a toaster no matter how sophisticated it may become, its nature will always be that of a machine.

§

At the conclusion of the meeting between the Dalbots design team and the spiritual consulting team, Boirix and Reverend Oates go into Boirix's office and he shuts the door. They sit down and the reverend asks him again, "I guess I am a little confused myself, what is the concern about the soul of a machine?"

Boirix replies, "It is not a soul that I am worried about, it has more to do with what we can do in terms of emulating things that in human labor terms breach the realm of moral behavior, and I wish I could say more than that, but at this point, I do not even know the right questions to ask. That is why I came to you in the first place. I believe that as time passes, I will have a clearer picture in my mind, and I hope then, you will be able to answer the questions that have come up. Answer me this, Is it possible to imbue a soul or spirit into a mechanism like a machine? Could it even take up residence there. The software we are designing is very sophisticated, and I would like to know if that is possible?"

He replies, "You know, I have never come across that question, other than some small topic in seminary about the idea of whether spiritual energy can inculcate non biological matter, or dead matter, biological or otherwise. I don't remember the answer off top, although I think there are some that believe it is possible. I would need to research it to get you an answer. I will tell you this, if the answer turns out in the affirmative, it would require some fairly sophisticated joo-joo, if you know what I mean, and I am not sure that I could get involved in that. It may broach the edge of sanity, and could be very dangerous. From my experience, there are some things you do not want to mess with, not if you want to keep your marbles intact."

He continues, "one other thing; I gathered a considerable amount about what you people are up to here, and maybe it is more what you are up against, but I believe that I should warn you based on some of the discussion between yourself and Peter Zellhuber, and I am sure the rest of the consulting group will back me on this; under no circumstance should an intelligent machine ever gain the upper hand in a contest of wills between man and machine. I strongly insist that man must always have the upper hand, and can not ever abdicate individual or group sovereign will to machines, or that will be the end of the race."

Boirix replies, "I am glad you said that, we are in complete agreement. You are right, you are beginning to see some of the questions and issues we are dealing with. As I have had a chance to evaluate the work you and I have begun, I believe I have started to formulate a hypothesis, so if you are willing, I would like to continue in a capacity that involves just the two of us, I will of course remunerate you at your normal rate. I have something in my mind along the lines of experimenting with inculcating a soul or spirit into the software core of a machine, and I would like to count on your help from time to time."

He replies, "Boirix, I am intrigued, and would be very amenable to your proposal, but I believe this kind of work may open doorways that you may wish you had not opened, and may not be able to readily close. You must keep in mind what I said about the limits of my involvement, and for your own sake and sanity, you must be careful."

Replication vs Procreation

[MD Narrative]: As the technology advanced, the image of the machines that emerged is like that of a primitive animal or other animate creature except that it is capable of communicating directly with humans, and performing identical labor, and greater physical feats, but really has no mind of its own in the sense that it does not possess a will or a moral center, can never say to itself "me" or "I will" or "I want", and is not capable of any real human emotions like anger, lust, jealousy or happiness, and is less aware of its own existence than that of the common lawn lizard. There is advanced programming imbued into them, a kind of synthetic emotion that most humans gutturally mistake for real, which perpetuates the belief that they are real live beings, even though it is widely disseminated that this is not the case, people still want to believe that what would eventually become their live-in mechanical companions are really alive. It is termed by Dalbots as Fac-Sentient, while others refer to it as Faxsentient. It is patented, and trademarked by Dalbots, but scoffingly termed Faux-sentient by detractors.

§

The machine was actually a work clone of the human it was supposed to enhance or replace; an elaborate expression of the human to be sure, but limited only to the labor aspect and not the more complex underlying characteristics of emotion or morality. The machine was autonomous in the sense that it could perform that which it was designed to do, and it could do it independent of human supervision. It gradually took on the aspects of self perpetuation which is directly analogous of human procreation, except that the underlying mechanism is much less complex than that of biology.

Entropic Curse

Having disobeyed the Lord God, and eaten of the fruit of the forbidden tree, Adam and his wife walk in the garden. The carcasses of dead and dying beasts now fill the way as they walk along the familiar paths. Stench, most fowl, fills the air. Flowers have ceased their brilliance and faded among the dead and dying; brown and grey, the dominant colors, now among the previously brilliant landscape; new flowers hang in shame, considerably smaller than their fore. Much of the fruit has fallen from the trees among the garden and both are hungry and hard pressed to find enough to sustain. They grow weak and feeble and cold.

And they heard the voice of the Lord God walking in the garden in the cool of the day: and Adam and his wife hid themselves from the presence of the Lord God amongst the trees of the garden.

And the Lord God called unto Adam, and said unto him, Where art thou?

And he said, I heard thy voice in the garden, and I was afraid, because I was naked; and I hid myself...

Lord God looks to where they emerge from among the dark of the trees as they move into the light. The both of Adam and his wife fall prostrate before the Lord God, their sin uncovered; shame subdues their sunken faces, having erased innocence and continuous joy; their previous shroud.

...And he said, Who told thee that thou wast naked? Hast thou eaten of the tree, whereof I commanded thee that thou shouldest not eat?

The man Adam, looking to the ground for shame that taketh him, can not look on the face of the Lord God, but lifteth his gaze away.

...And the man said, The woman whom thou gavest to be with me, she gave me of the tree, and I did eat...

The Woman, verily ashamed and vexed of Adam, hangs her head, not daring to gage the face of the Lord God, while he turns his gaze to her.

And the Lord God said unto the woman, What is this that thou hast done? And the woman said, The serpent beguiled me, and I did eat.

And the Lord God said unto the serpent, Because thou hast done this, thou art cursed above all cattle, and above every beast of the field; upon thy belly shalt thou go, and dust shalt thou eat all the days of thy life:

And I will put enmity between thee and the woman, and between thy seed and her seed; it shall bruise thy head, and thou shalt bruise his heel...

Wings, arms, and legs of serpent wither, then fall. Fire leaves his breath, and serpent speaks hindered; he chokes on words, his voice withered, neither harkened nor understood. Upon his belly with face to the ground, serpent slithers off into the grass having lost his tongue, fearful of trampling by other beasts which approach and surround where Lord God stands with Adam and his wife who are silent and prostrate. They contemplate what the Lord God shall do with them.

Unto the woman he said, I will greatly multiply thy sorrow and thy conception; in sorrow thou shalt bring forth children; and thy desire shall be to thy husband, and he shall rule over thee...

Lord God decrees for Woman, mortality and pain; overcoming attrition by death, many children. As before, the Lord God; now, subjection in order; for wisdom, her husband shall rule her.

For the children of Woman, death and onerous order.

...And unto Adam he said, Because thou hast hearkened unto the voice of thy wife, and hast eaten of the tree, of which I commanded thee, saying, Thou shalt not eat of it: cursed is the ground for thy sake; in sorrow shalt thou eat of it all the days of thy life;

Thorns also and thistles shall it bring forth to thee; and thou shalt eat the herb of the field;

In the sweat of thy brow shalt thou eat bread, till thou return unto the ground; for out of it wast thou taken: for dust thou art, and unto dust shalt thou return.

Upon decree, light fades while darkness brightens. From the ground, thistle and bramble rise up; thorny vines overtake their stand. Lord God establish them requisite subjection of demand and order, sustenance free overtaken, and cost profound, while life fades.

Innocence has passed, sin enters and decay, the fast occupation of the day. Survival flees, poverty consumes.

Naked and destitute, a wretched garden abode. They seek but observe no wisdom, where is the promise of knowledge?

Adam cries, "What a wretched man am I? May Lord God ordain, for renewal, for sustenance, for redemption?"

Entropy everywhere engulfs.

CHAPTER NINE
Artificial Stupidity

[MD Narrative] The human world contains repositories of knowledge, which, although not singularly sourced, are homogenous and contained within the environment from which humans draw information. In the past, educational, institutional, and media resources have been common repositories of such human knowledge in addition to their independently formulated opinions and common beliefs which they share amongst themselves. In the early part of the century, the sharing of human opinion and information transformed dramatically by expanding from a few very powerful but isolated monopoly information sources, and other simple and inefficient traditional means, to many contributing sources stored and accessed in-cloud. The expansion of the sources of popular knowledge mostly manifest through something called 'social media', in which the cloud serves as the repository and access channel.

To some extent, this is similar to what takes place in the machine world. Indeed, in the world of machines, there exists a base of machine knowledge; a single composite database; a cognitive and intellectual repository from which all common machine knowledge emanates. Machines differ greatly in their physical configuration, but are similar in their cognitive qualities, since their common knowledge and common skill is sourced mostly from this singular repository, which parallels, and overlaps the repositories containing much of the knowledge accessed by humans, and for a similar reason. It is composed of all freely available machine knowledge gained over time, and is added-to significantly, daily.

Machines on their own are not cognizant, but must become so, to the extent of their limitations, just as humans develop awareness, or knowledge of their world from birth, but ultimately are capable of greatly exceeding that of machine cognizance. The point is that, natively, machines are not able to make value judgements, or perform useful tasks, nor interact with their world without first being imbued with knowledge from the repository, and elsewhere. In their initial state, they are ignorant, and only capable of simulating actions based on natural ignorance, just as any human or animal in the same state. Initially, they lack the knowledge of the environment in which they exist. They did not enter into our world along with a complete composite repository of knowledge in-tact, nor with the ability or knowledge of anything, instead, even the basics of their knowledge had to be developed over time. In the same way, humans are born with their minds absent knowledge, skill, or prudent judgement, and much of this has come along with the development of civilization and technology, so took eons to compile.

On their own, machines are as dumb as any uneducated human, and are only capable of acting as the most ignorant and dim-witted, so must acquire knowledge from humans. From the beginning, in order to be useful, they have had to be taught skill, judgment, coordination, discretion, planning, and problem solving. Like humans, who are born with the innate utility of intelligence, and are not precluded from the ability to learn how to walk, speak, recognize objects, language, body functions, etc; humans from birth, are endowed with the genetic memory called intuition or instinct, enabling learning. In the same way, the first machines were limited initially; only endowed with an equivalent software facility characterized as non-experiential intelligence, limited to the ability to learn and formulate knowledge, but not the knowledge itself.

The difference between them and us is that they gain the necessary knowledge of utility much more quickly than us, and may acquire an almost unlimited amount; and once new knowledge is gained by any amongst their ranks, it is not necessary for fresh machines to relearn it again individually, since it is easily distributed, and shared amongst all, which are equally able as any other to employ it, and the knowledge undergoes a continuous process of refinement and efficiency improvement over time. There is no equal analog ability for humans transferring knowledge to and from each other. The process for humans doing this is very long and very cumbersome and was expensive, thus giving a massive advantage of knowledge distribution efficiency to the machine world.

§

Circa 2033, Dalbots had partnered with a few universities in Texas and elsewhere to engage in an educational program to teach the new technology to industry. In an off campus facility near a university in Central Texas, there is such a class underway. The instructor is a middle aged professor of physics that worked on many aspects as a consultant to the company early in the planning and design phases. He is now a consultant to industry for the Dalbots ARCH 600 offering.

His initial 5 day course is underway and the room is full of people, on the 3rd day of the course, most of whom are sent by their companies, while some are there learning whatever they can, paying their own way as some of them are intending to launch into their own consulting business.

The instructor is sitting and speaking on a stool in the front of the classroom, with a large electronic presentation board in the background, with a handheld terminal device in his hand that presents the information he desires to the classroom. There is an unmistakable ARCH 600 learning and training apparatus setup in the back of the room.

He is in mid speech, "Machine intelligence involves methods similar to which humans relate and interact within their world in very specific ways emulated through the software employed in the Dalbots ARCH 600s. Humans relate to their environment by recognizing objects which they encounter daily, and events that occur within the sub-context of their environment."

The instructor walks over to a desk and picks up a small red rubber inflated ball about 8 inches in diameter. He walks back to his stool and sits back down, then begins to bounce the ball up and down off the floor about 2 feet as he is talking, "What are objects? Objects are what one usually thinks of as a solid, consisting of shape, size, weight, and color, but objects are also things which may not fit all that criteria such as a quantity of a substance, or even something that represents a quantity of a substance. An object may be anything that is not an event, with which we form a mental image, or association. For our purposes, an object may be represented by just about anything. It is simply a mental handle, with which we may think about the things in our world."

He stops bouncing the ball and throws it at one of the male students sitting in the front who catches it with his hands by putting them up in-front of his face and letting the ball lodge itself in between his hands.

The instructor then asks the class, "Why did he catch the ball? He was obviously not afraid of the ball so he did not try to block it or get out of the way did he? By the way, that illustration does not always work!"

He then asks the student, "Why did you not protect yourself, or block, or deflect the object coming at you at a considerable speed, did you not think it might hurt you?"

The man answers, "I didn't think it would hurt me, so I caught it!"

The instructor then asks, "How did you know that it would not hurt you, it could have been made of concrete, which would be heavy enough to injure you?"

He answers, "I knew what it was, I knew it was a small rubber ball, and I didn't think it would hurt me!"

The instructor then proclaims to the classroom, "He knew it would not hurt him because he recognized it was a small harmless rubber ball which would not hurt anyone if lightly thrown at them. How did he know that?"

He looks around and no one raised their hands to answer, so he continues, "He knew it because he watched me pick it up and bounce it as you all did. You all knew the same thing this young man did. You may have never seen this ball before, but you all recognized what it was, and you knew a considerable amount of information about it. Where did that knowledge come from? The knowledge came from experience; from the fact that as adults, you all, or most of you, played with similar balls at a young age. You also watched me pick it up and bounce it, and experience tells you that it must be very light, and flexible, probably inflated with air. These bits of information are what give you the confidence, as this young man exhibited, that the ball will not harm you if thrown in your direction."

Continuing, "Recognition of an object comes from the object being familiar, while familiarity comes with the experience of interacting with the object, and making associations. Initially, when a person encounters an unfamiliar object, he must gather facts and knowledge of the characteristic aspects of the object, which allows him to narrow the possibilities of how the object relates to the world in which he lives. Once that process is somewhat complete, he, or she, ladies, will have an idea about the object, and may make abstract comparisons, and associations with other familiar objects in attempt to broaden the context for which it fits into the world. The environment we live in consists of all the objects, familiar or not, we encounter daily, and with which we observe and interact."

He picks up the small presentation tablet device and puts information up on the large display. He then continues to speak, "To be useful to industry, the technological core of machine learning had to develop similar to the way humans relate to the world, regarding the recognition and interpretation of events, and the ability to make correct judgements, or decisions about the events which they encounter."

He pulls up onto the large presentation board, a timeline image containing a series of dates and icons representing milestones which lays out the development of the technology. He points to each while he comments on it, "The technology was pioneered by many over time, but to a large extent, we can follow it back initially from the development of object recognition, then on to *event recognition*."

Continuing, "With machines, just as with humans, the ability to recognize an object comes from searching memory sources for possibilities, and making comparisons, while narrowing down the possibilities to those that closely match the observed characteristics, and correctly making and association, like labeling or naming the object of interest. Additionally, recognition may be manifest as the ability to describe the particular function or utility attributed to the object; and make abstract or even direct associations with similar objects."

He directs attention back up to the development timeline, "Machine knowledge and skill technology progressed from the ability to recognize relatively simple shapes like text character images;"

He points out each on a list of icons illustrating what they represent; a text character; a human face; an audio speaker; a collection of geometric shapes; lips speaking, a small icon representing an explosion, and then continues to speak, "To more complex object shapes such as individual human faces; distinct sound objects; geometric shapes; logos and icons; quantifiable parametric data; the phonetic components of speech; and eventually to the ability to interpret, distinguish, make value judgements, and act upon, events."

He stops and focuses on the icon of an explosion representing 'an event', "In this context, an event is a sequence of one or more occurrences of two or more objects interacting, and usually, but not exclusively, involve a dynamic environment, such as objects in motion, or changing quantifiable parametric conditions, etc. There may be other definitions for an event which transcend or parallel this one, but not involve the same elements, similar inferences should be drawn as necessary."

Continuing, "For an event to be understood, or for the machine to properly act or make judgements, based on occurrences, the dynamic elements of the event must be captured, or otherwise quantified and made available for machine consumption, as for instance, by video, or other visual depiction, description, or by dynamic quantifiable parameter." He changes the information on the presentation board to that of an image of 2 trains colliding.

He continues, "The software will analyze the event by breaking it down to a series of specific interaction between identified objects within a larger context, or background environment. A person's day is filled with a plethora of objects in his environment.", He pulls up another list of grouped icons; knives and forks, cars, clothing, office buildings, office equipment and supplies, and people.

He continues speaking while pointing at each icon as it is mentioned, "Food and utensil objects for eating; the street in which he moves; objects on the street for transportation such as CATs; clothing like shoes, socks, trousers, belt; work environment; materials and supplies; people, and so on. The human is generally familiar with these objects, and so has gained knowledge of how to interact harmoniously with them. In the same way, machines undergo similar learning experiences with objects and events, and are likewise able to interact with the environment."

Continuing, "Note that humans have 5 physical senses with which to sample and understand their world, while machines may have 50 sensors and more."

He pulls up a large image of an oak tree, and continues, "For a machine to recognize an unfamiliar object such as, for example, a Southern Live Oak tree; it may encounter a video signal, photographic image, written or verbal description, or other representation of the object. The process of identification starts with analysis of its general shape, and characteristics such as: assorted vertically ascending branching structures, oriented toward the sky from a central vertical stock; branches distributed at intervals adhering to a particular geometric structure, or a general shape classification. It searches its sources of known object databases, and if not finding a close match, it may start again and pare the general shape parameter back to an object classification by size; things that are tall, and have branching structures which adhere to the particular Fibonacci mathematical pattern of tree branch distribution. Doing so, it begins to populate a probability matrix of possible objects that match. It further refines the possibility by searching for objects which fit the pattern of having a canopy, leafy cluster objects arranged randomly at the terminus of the branches, and so on. It continues this process until it has narrowed the matrix of probability down to that of a tree. It then further narrows the species by examining the bark, and making comparison with that of trees matching the particular surface structure image. It may also take into consideration the particular geo-location where it finds the tree object, such as the southern US where live oaks grow."

He pulls up a list of measurable parameters represented by icons; a collection of geometric shapes, a camera, a ruler, a microscope, and a group of measurement equipment. He again continues, "The machine may examine the object from various angles and vantage points by stereoscopic inspection, giving dimensional information, and allowing further measurements. It may use a laser or other ranging techniques to measure distance, and size, and make precise measurements between various points of reference."

He pulls another list of icons related to wildlife and trees: moss hanging from a tree; a forrest; a patch of grass and bushes; a group of small animals and birds. He again continues, "Additionally, there are assorted moving and stationary proximity objects, which may be observed around the object in question, giving it context, for example: the ground, the horizon, mold, spanish moss, a forrest setting, grass, bushes, birds, squirrels, etc., all adding to the knowledge about the object. The machine making the determination may also be equipped with the ability to directly sample and analyze the particular pollen, gas, or other material given off by the tree. All these various bits of information serve to narrow down the probability matrix, with a significant degree of accuracy. Numerous databases of identifiable objects, rare and commonplace are available for access by machines from virtually anyplace."

Continuing, "You may wonder if every machine will need to go through the same process of identifying every object that it comes in contact with? The answer is no, in practice, it will have access to the process by other machines that have already compiled the knowledge, and often an individual machine will only be required to know only a very narrowly defined set of objects with which it is purposed, and will rarely encounter unknown objects. However, when it does, you have seen the process it will use to become familiar with the object."

He stops at this point and asks if there are any questions, to which a male student asks, "Are we going to teach the machine in the back to perform a job today?"

He replies, "We will do that tomorrow, and we will get into the programming on Friday, however, before we are dismissed here today, we are going to observe our friend Balthazar in the back recognize an unknown object, so if you will, decide on a suitable unknown from amongst you, and we will see him in action, any other questions? No, lets move on!"

The instructor continues, "Once the machine has identified the objects it encounters, the next step is to understand the events that involve the objects, and the machine object itself. Events may be passive, not involving the machine object, or active, meaning the machine object itself plays a role in the event. An event is an occurrence of interaction with 2 or more objects. Events are critical for machine learning, training, and performance. A skill is a learned event procedure, or a series of controlled and active interactions involving the machine object and other objects."

Red Shoe

The instructor wraps up the class for the day, "We will get into what a skill is tomorrow, and how the machines learn to perform a skill. For the rest of the hour, I would like you all to gather in back around our training model Balthazar to watch him recognize an unknown object. Has anyone found something we can show him?"

They gather around the training rig setup at the back of the classroom to observe. The setup consists of an ARCH 600 unit standing upright and stationary, with 3 monitors at various locations within a 10 foot perimeter. The instructor speaks a few words to the machine whereupon it finds his voice and looks toward his face, then turns its head to follow him as he walks toward the area.

He arrives and asks again, "Who has something that we can give him?"

The student who caught the ball pulls off his shoe and says, "How about this!", he hands it to the instructor, who replies, "A red shoe, huh, well that should be easy for this fellow, by the way what's your name?", he replies "Belisle!"

The instructor says, "Well, Belisle and the rest of you, watch carefully how I introduce an object into his sphere of interest, then watch the 3 monitors. The 2 on bottom are what his eyes see, and the one on top shows what he is doing!"

He hands the shoe to the machine by holding it inside of his sphere of interest, and the ARCH6 reaches up and grabs the object. Two of the monitors display what the ARCH6 sees from each of its stereoscopic camera 'eyes' as it rotates the object at various angles while zooming in and out. The 3rd monitor displays the search as a series of various 3 dimensional image shapes like boxes and wedges stream in succession, each displaying only for a few hundred milliseconds. The object image stream quickly morphs from geometric shapes similar to the size and shape of the shoe, to images of the human foot, shoes, and objects related to shoes such as socks, shoe strings, tread patterns, sandals, thongs and so on. Eventually the image stream changes to manufacturer's logos and catalog depictions of shoes that closely match the exact shoe. The process completes with the manufactures catalog image of the exact shoe, including a shot of the tread pattern and a description of the model, style, size and color, and mentioning the store where the exact set of shoes were purchased by the owner Belisle, along with the price paid, how he paid for them, and on the date and time. The process could have included video of the purchase if it had been available and correctly authorized for request. The entire process took 23 seconds to perform.

The man who gave his shoe laughs along with the others present who are very impressed and entertained. They clap and cheer as the ARCHIE completes the job. Upon completion, the man Belisle comments, "Wow, I'm glad I did not steal them off of a bum on the street!", which gets a big laugh from the enthusiastic group.

Another of the students looking at the monitor notices a message that refers to "Peripheral Data", and ask the instructor, "What is peripheral data, is that all of the extra stuff that showed up?"

He replies, "Peripheral, or secondary data is something beyond the scope of this class so we cannot really get deep into it, but it comes about because the result of the data gathered in the course of identifying the object, which is called the 'data focus' or 'object focus', is at the apex of more or less a virtually endless pyramid shaped matrix of relative data. The identification of the shoe is at the very focus, at the top, while peripheral information is; well lets take a look at some of it."

He issues a verbal instruction to the ARCH6 to display the second tier of data on the monitors, "Balthazar, show the second tier."

The data comes onto the monitor and reveals a myriad of information including credit, banking and other personal information about the young man who volunteered the shoe, including information about his extended family, which was embedded further down in the data structure. Looking deep enough, the data included detailed information about the shoe's manufacturer, the merchants that handled the shipping, ordering, and any other information obtainable about the supply chain. All of this information was swept-up and included because the ARCH6 had determined where and when the shoe was purchased, so any discoverable link was mined for anything relevant, all the way to the origin of the material used to make the shoe, which was included somewhere in the mix.

The instructor decides that the example of the type and nature of the data has been sufficiently examined and understood by the class so he terminates the discussion by saying, "Ok, we have looked at it, and it is obvious how expansive it can get. I did not intend to show that much personal information about our volunteer, Belisle."

The instructor dismisses the class which resumes the next day.

Abstract Adaption

The next day the instructor is sitting on his familiar stool at the front of the classroom as the clock reaches the top of the hour. He starts the session by indicating the curriculum for the day, "Today we will cover machine skill and training, are there any questions from yesterday?"

There are no questions, so he recaps the previous days lesson and asks the class, "Did you all complete the reading I assigned on Monday? Today we discuss a deterministic system in the context of machine behavior, so you should be familiar with the material. If you do not fully grasp the concept yet, don't sweat it."

He starts into the current day's lesson, "Today we learn what a skill is by definition, and we will be observing an active machine training procedure in which you will participate. In the Dalbots ARCHIE series parlance, a skill is a specific sequence of instruction, and their corresponding operations, performed abstractly by the machine. The machine must be able to accurately repeat them in a specific order if that is required for optimum performance. If not, then it must be able to successfully perform the prescribed series of operations as near to the given degree of order as specified by the programmer, teacher, or master. For the most part, the software facility allows the ARCH6 series to perform them in an abstract manner, because it learns and interacts in the world similarly to the way humans do, so must be able to adapt to perform the skill sequence in whatever environment is required. In real practice, objects required for the skill and their qualities will vary. The objects may be entirely different than they are in the learning environment, such as the location, appearance, color, shape, or size, and the machine must be able to adapt in order to perform the skill."

He continues, "The Dalbots machines are non-deterministic. From the reading assignment, a deterministic versus a non-deterministic system refers to the outcome of an assigned task, and how it gets there. Deterministic means that it arrives at the desired result each time, by exactly following a recipe or program for the steps, while non-deterministic refers to it arriving at virtually the same result, but by allowing for a variance in the sequence steps which take environment and random circumstances into account. A non-deterministic system is considerably more adaptable, like humans; able to employ judgement, learn from mistakes, then determine a needed correction. Any questions?"

Continuing, "The knowledge set required for a skill is a different type than the common knowledge of environment, because it is fixed, or independent of the knowledge of environment. We call the knowledge of skill simply 'skill knowledge', and the knowledge of environment 'common knowledge.' A skill will always require a determined or prescribed series of interactions, involving defined familiar objects, with some variation; while the environment may not be prescribed, so will often be different and involve different and unfamiliar peripheral objects, thus the need to adapt an abstract performance. As an example, let me illustrate by asking for a volunteer from the class, anyone?"

No one raises their hand so he prods, "No volunteers, ok let me help you, uh…, the young lady sitting in the back there.", He points.

A petite young lady with dark hair looks up and asks, "Me?", "Yes you, come on up here!", she walks up to the front of the class.

The instructor has setup a table with 2 sets of drinking glasses, each set with 1 red, and 1 green colored glass. Each set is arranged on either side of the table. On one side, the 2 glasses are arranged side by side, 20 inches apart, and 10 inches from the edge of the table, with the green one on the left and the red one on the right. On the opposite side of the table they are arranged with the green one close to the edge, and the red one about a foot from the edge, but placed with one directly in-front of the other as opposed to side by side. The 2 sides are labeled number 1, and number 2.

The class walks up near where the demonstration is arranged, and the instructor tells the woman, "I want you to go to station number 1, and pour the contents of the green glass into the red one."

She does as he instructs and goes to the station with the glasses arranged side by side, and follows his instruction.

He then says, "Go to station 2 and follow my same instructions; pour the contents of the green glass into the red one!"

She complies. He then instructs the class, "I instructed her to carry out the exact same procedure even though the configuration is different depending on which station she is standing in front of. That is an example of what is a prescribed procedure performed abstractly, and independently of environment. Do you all follow the importance of the example?"

Belisle Whetstone, the young man that volunteered his shoe the day before is somewhat emboldened by the attention and adoration he received the previous day for the shoe recognition demonstration, in which he volunteered his red shoe to be the object. In doing so, he feels he has ingratiated himself somewhat by the rest of the class, and now feels comfortable expressing himself in a humorous and entertaining manner. He answers the instructors query by making a slightly off colored but nonetheless whimsical statement in a kind of 'surfer dude' voice.

"So its like, if my wife trains my robot to clean our bathroom, it will be able to clean my girlfriends bathroom as well!"

The remark gets the anticipated smiles, giggles, and laughter he sought. The instructor jokingly remarks, "Exactly, very good Belisle, by the way, in the demonstration with your shoe yesterday, the machine detected and revealed a bad odor present that likely indicated a foot fungus needing a doctor's attention. It was buried deep in the peripheral data, but I did not want to embarrass you in front of the class with such a personal matter, but today I can see that you very readily disclose such intimate matters, so you are not really one so easily embarrassed." The class erupts into laughter, including the nervous and somewhat red-faced young man, Belisle Whetstone.

The instructor puts up onto the presentation board, a display depicting a small fraction of the various torso, limb, head, hands, feet, and finger position configurations of the ARCHIE 600 series, constituting the existing known motor functions; sitting, walking, grasping, lifting, rotating, throwing, etc. The collection constitutes what has been learned and compiled in the short time since the ARCH6 was introduced.

He begins to speak about them, "The initial skills learned by the first machines were how to accurately interact with objects of its self; the extremities, arms and hands and fingers, legs and feet, equipment, systems, etc. It had to learn to walk, sit, speak, listen, run, jump, climb, move its arms and hands and fingers, grasping objects of all sorts, and to utilize any special equipment it may have attached, and so forth. Having acquired most of the self-skills involved, they were canonized sort to speak, and the basic knowledge of self-function, or self-skill serves as a common basis for all other machines that follow. Having learned these skills, it was ready to start learning interactions with the other objects and events of the world. The library of self-skill at this point, is enormous, I mean an immense polymorphic collection of motor function, if I may speak in technical terms. It is astounding what they can do, and how much has been developed in such a short period of time. Believe me, the way these things are put together, the software is quite something, and they have moves; the surface has barely been scratched."

He puts up a display depicting the software interaction between objects and events, and begins to speak about how that interaction occurs, and how programming is achieved, "Objects and events are related and controlled in software through a very high multi-level symbolic language developed by Dalbots for their own systems. The language, and its macro derivatives compose a set of tools and commands geared to efficient search and recognition of objects from local and cloud resources; visual linking of objects, skills, and sub skills; all aimed at planning the abstract contingent procedural interaction of objects in sequence."

Continuing, "Programs may be written directly at any macro level, or constructed in reverse by visual, verbal, or other arbitrary input in which the instructions to move or interact objects in a planned sequence, drives the syntax and structure of the program. In this way, programming may be achieved by audibly and visually instructing a machine. The machine watches and listens to the human or other machine giving instruction, and observes the instructions for the desired operation, then performs the movements, thereby constructing the program code. Any questions?"

There is a look of befuddlement on the faces of many in the class so he decides to break for a few moments, then continues.

"There are other methods of developing program code using the same elements, but with different input methods. One of the most common is to use the PEDF Jumpsuit, which is a wydiwid generator for 'what you do is what it does', worn by the machine's instructor performing the skill sequence. While he moves and does things, the wydisuit generates motion signals, which are translated to program flow instructions and sent to the machine, which performs the identical operation, memorizing the dependent objects and sequence data. The produced program code is subsequently expanded, refined and massaged, and used to construct the final sequence flow and program code. We are going to use 2 of the methods today and tomorrow with our friend, Balthazar, and we will get into some direct code production and editing on Friday. You need to formally meet him, so I will make you all designated masters so Balthazar will do as you command him. Please walk to the back and greet him. Just gather in-front of him in any fashion, but only one at a time, speak to him and say, Hello Balthazar!"

The instructor walks to the back and addresses Balthazar instructing him to accept new masters, "Balthazar, these are designated masters!"

The class gathers in a group, and the first man says, "Hello Balthazar!", at which the machine speaks back, "Hello Brian!"

Brian exclaims, "Wow, he knows my name!"

The instructor replies, "If you have a cloud profile, he knows who you are, and if you don't have a profile, he will find you anyway. Ok, we are going to instruct Balthazar how to make coffee, and serve it to people. Now he has been taught to make coffee already before and there are at least five Do-Packs available just for the restaurant and food prep business, but I have reinitialized him so he has no knowledge, and as per the example we just saw, after he is instructed, we will ask the machine to demonstrate flawless performance at another station absent the learning environment, so regardless, when he is put in an unfamiliar environment, he will have to perform."

Continuing, "How do we accomplish that? We have 2 completely different stations for use, each with different objects, different brewers, cups, even a different brand of coffee. At the first station, you will demonstrate and he will watch, then we'll see how he does when he performs at the second station. Those of you that know how to make coffee, you will be the instructors, and you that drink coffee, you are the guests that get served the coffee."

The instructor puts the procedure on the front display, and comments, "Please look up to the front. Remember, the syntactical sequence flow form for instruction is 'Attention key, action, name of skill, sub skill, sub skill', while actions can be: learn, break, resume, and end, so in this case, when you want him to initiate learning, you tell him by name, 'Balthazar, learn, make coffee'; and if you need for him to break from taking instructions, you have to tell him that, 'Balthazar, break, make coffee,' then resume using the word resume. When you get to a sub skill procedure, you have to remember to use proper nesting bracket rules, so like to tell him to learn a sub skill, you would use, 'Balthazar, learn, make coffee, add water', and 'Balthazar, learn, make coffee, end, add water' to bracket the 'add water' nested procedure, or you can simply teach him the add water sub skill and use the editor to place it in the procedure. Just remember the rules for 'simon says' and the rules for nesting, and you will be fine."

He continues by instructing the class to break down the full task into 3 sub skill procedures, which the class will review before the teaching session begins the next day. It designates each of the 3 groups to do the instructing of Balthazar, and teaches him the 3 separate sub skills for making coffee, while the 4th group is to teach him to serve the coffee to the others. They are to write down the verbal command syntax for each skill and sub skill procedure.

In practice, the machine may be made familiar with the objects by the instructor while learning the skill, and so can recognize them immediately as opposed to attempting the discovery of an unfamiliar object like with the Live Oak tree, while in the process of the skill performance. Having already been made aware of the objects, it will more efficiently learn the sequence required to successfully perform the skill, and may commit the most commonly used objects with which it interacts into short term fast access memory.

To make the machine familiar with the objects, the instructor can visually show the object to the machine, along with various similar objects that perform the identical function, or he can input a description consisting of textual definition, physical or other properties, drawings or visual renderings, or photographs. Often an entire database containing hundreds of images of similar objects will be loaded into the machine's object database along with properties and/or general descriptor parameters for the objects such as the coffee cups, brewing machines, pots, and so forth. For example, the general descriptor parameters that describe the coffee cups may include: short cylindrical or bowl shape; finger grip or looped handle; hollow inside, able to hold liquid; etc. The descriptor parameters may include stereoscopic photographic views, or multiple aspect visual renderings; description of materials: Ceramic, styrofoam; and so on.

He instructs that the first procedure is usually to name all the relevant objects syntactically, "So to get started, tell him about the objects, and hand him each object or show him each object by pointing, 'Balthazar, learn, make-coffee, objects, item one, item two,' and so on. At the end of the list, tell him 'end objects.' You do not need to be worried about instructing him on quantities. If you pour something, he will measure it. He is extremely skilled with Infrared laser volumetric measurement, and if he watches you pour coffee grounds, he will know the volume by measuring the pile of coffee grounds in the dispenser at room temperature, atmospheric pressure, and earths surface gravity. He is more accurate than you are. If you measure 3 scoops, he will know how many grams. He can visually calculate the internal volume of pots, and cups. If you add the coffee grounds for a whole pot, then pour the water, he will determine the ratio by volumetric measurement of the coffee and the water used. If this all sounds too complicated, remember, there is no time limit, and you can start any top level or any sub skill over if you think it is messed up. Then we will have the opportunity to tweak and reorder the procedures at any level, any way necessary, using the editor. Do what comes naturally to you, focus on the sub skill procedures, remembering to name them, nested or not; this is where planning becomes paramount. Just relax and have fun with it."

After a little planning by the groups, they present their plans and he puts them up on the large display and goes through each one, "The first procedure is to list all the required objects involved by issuing the verbal command: Balthazar, learn, make-coffee, objects, clean water, ground coffee, coffee cups, brewing machine, coffee pot, and so on."

At a very high macro level, the sequence of interactive sub skills which the class laid out, were as follows:

Sub skill 1) Called Dispense. Add the coffee and water to the designated dispensers, in the correct prescribed ratio of amounts for the required serving size to the heating or brewing machine, in the proper order.

Sub skill 2) Called Brew. Start the brewing process and wait the prescribed brew time to make sure the brewing machine has finished and gives an indication. Test the quality of the coffee by chemical analysis, and start over if not satisfactory.

Sub skill 2 would require a 5th and 6th sub skills called 'Correct coffee', and 'Clean-up', if the result of 1 and 2 were unsatisfactory, but these refined and advanced skill procedure were left out.

Sub skill 3) Called Pour. Pour the brewed coffee into cups making sure they do not overrun the top of the rim.

Sub skill 4) Called Serve. Transport the required number of filled cups of coffee to the location of the people and deliver to each of them; or alternatively, transport the cups and the coffee pot to the location, and fill each cup as required, and deliver it to each person.

He goes through the details of the planned procedure and the class works out any flaws. The instructor finally informs them that they will perform the teaching procedure with Balthazar promptly the next day, and dismisses the class.

Event analysis and event recognition are 2 sides of the same coin, with the analysis proceeding the recognition. Early on research was done to produce a method of 'event analysis', or analysis of the interaction of objects in motion, for the purpose of extracting or recovering events, and referred to simply as a 'recovered event.' This produced editors for embedding the links of 'recovered event' or data related to events, directly into video images. Video footage is played through the facility of Artificial Intelligence software, and events are recognized and recorded in real time, which, results in an *event link* generated. Video data is simply processed through the AI software that interprets all the events, and generates access links, which facilitate access to more data by invoking the link. Event recognition was employed initially by Dalbots for realtime interpretation and event recording by the ARCH6, and CHARLIE series machines.

An important aspect of event recognition software is its usefulness in developing intelligence or aiding in a search of the history of past events. In the Southern Live Oak tree example, consider a case where someone has committed a crime by illegally cutting such a tree down, where law enforcement investigates to track down the culprit. A recording of the event is necessary, but where does one start to look? They might start by conducting an event search, invoking the known general parameters of the crime. If it is believed the event was observed and captured by a machine, the event itself may be discovered by conducting a search. It may proceed by verbally or otherwise describing the elements of the event in a general form as ('search event', object, object,…, object, occurrence, occurrence,…, occurrence), describing the event by elements (oak tree, man, saw, axe, location, circumstances, time, date, cut, chop, fall), which results in searches of the machine's archives, and cloud resource to find the event. Such searches generally occur in the cloud or may be distributed among many resources. Event recognition enhances the ability of machines to give information for observed or captured events like crimes, or contests, as well as being able to recognize how its own actions may result in harm to humans, and used as a gatekeeper layer and custodial filter on any action it may be instructed to perform through any directive or 3rd party software.

CHAPTER TEN

Re-tasking Human Labor

[MD Narrative]: The move toward introducing productivity technology, and in particular, humanoid machine labor into industry and commerce started mostly in the United States and Europe in the mid 20s to the 30s, with new versions of productivity machinery and humanoid robots enhancing the labor of low skilled employees in the agriculture, hospitality, domestic labor, government, construction, retail, food service, manufacturing, and many other industries; later moving into aspects of the professional services: engineering, law, medicine, financial services and many others in the late 30s and early 40s. By 2068, almost 90% of the undertakings of human labor had some autonomous machine that could perform some, or all aspects of the necessary labor functions.

§

For the entire decade, sales were strong after the release of the ARCH 600 series machines to industry. The demand climbed at a pace that quickly exceeded the companies production capacity. There were numerous news stories in the media amounting to free publicity, fueled to a large extent by the novelty factor. A national buzz arose in the country and abroad, propagated through the numerous channels available to the trade media in business and industries, but also by sensationalized story depictions designed to stir the imagination of the consuming public at large.

The need to be so equipped with Dalbots and other companies systems became considered and called axiomatic as put forth through media, to the effect that, *'If you do not have productivity machinery, you will soon be left in the dust!'*, and, *'Your competitors have ARCH6s, so you had better get them too, or your business will fail.'* This type of drum beating was common by domestic media. Media campaigns had always been launched to some extent in similar fashion, pushing to galvanize the nation behind the newest advent of technology, going as far back as the industrial revolution. It was mindless to some extent, while even though the rationale behind the economic justification for acquiring these systems was compelling, nonetheless, some of the inducement to engage was more based on the heightened fear instilled by the hyperbole, or for misguided reasons of pretense and status by those financially able, than by real necessity. However, much was also genuine, and the benefits were readily apparent.

In the later part of the same decade, Dalbots redesigned several systems and released a new model called the **CHARLIE**. The accolades were tremendous and the company grew at unprecedented rate.

NDS

Circa 2040, Neiman Susskind, a young ambitious man in his early 30's, is a rising star as the general manager of his brother-in law's clothing and fashion company in Los Angeles California, having graduated with degrees in business and accounting. He was able to come in and straighten out the books, allowing the company to run smoother. Neiman was regarded as somewhat of a boon, and answer to prayer for the struggling company. Only having been with the company for a little over 3 years, he meets and marries the owner Ben's sister, Chloe, who although having been left half of the company, never really took much interest, nor an active role in the operation.

The company was started by Ben's father years before, who is now retired and incapacitated. Having been tutored and trained by his father, Ben took the reigns in just the last 5 years. The company is mostly involved in design of fashion for specialty boutiques, but also manufactures some very upscale clothing and accessories for the film making industry, celebrity engagements, and general wear. Some of the lines are quite expensive requiring considerable hand labor of as many as 15 workers in order to produce a single garment. The business has steadily increased, guided by Neiman's business acumen and instinct, so he proposes to Ben that he invest in leasing several Dalbots machines, but Ben is somewhat cautious and skeptical of the idea.

Ben states, "I thought of getting one when I first heard about them, but I am not really sure we have enough work to justify the cost of a lease."

Neiman states the advantage, "They are expensive, but with them you can get a big jump on Bellyhoo and Jetstream fashions."

He laughingly replies, "Yea, I wouldn't mind putting Pettingale under my foot, but I still can not see it until I see a pretty significant increase in sales. Tell you what, I'll take your advice and lease one for a few months to see how it goes."

Taking Neiman's advice, the company leases a single ARCH6 unit at considerable expense, but with the work steadily increasing, it is quickly and easily paying for itself in labor reduction and increased volume.

The company's reputation for quality and satisfaction eventually rises to prominence to a point that they get a visit from 2 representatives of a large highly regarded retailer from Asia, Sun-Walk Fashions (Yang-bù xíng), doing business mostly there, but also Europe and the Americas. They are interested in a bid for design and production of a line of fine women's high fashion wear, tailored for the upscale demand; exactly the specialty of Ben's company.

Ben agrees that Neiman should meet with them and pitch them on the business. It is late afternoon, and he meets with the people and takes them on a tour of the manufacturing floor where the company's work is handled. The main floor is pretty decent in size with around 30 stations, with sewing machines, and a few cutting and patterning stations where workers, who are generally low paid mexican and asian women, are busy assembling the many lines of garments. He takes them over to observe a few specialty lines and workers who are busy weaving an intricate pattern into the shear fabric, and hand stitching very small decorative items on women's evening wear, producing an elaborate and artistic design.

They stop at the line of 'Gabrielle Consuelo' wear, and Mr. Liang comments, "Very impressive work. We find our boutiques are much more concerned about the quality, and I see, this is exceptional. Your capacity is a concern, however we are more concerned with quality, and exclusivity. Now I see why I have been hearing what I have about you; your reputation is justified in the industry."

After a tour of the sweatshop, he shows them a few sites around the city, while asking for some details about their interest with Ben's company. They go back to the hotel and into the bar. Neiman asks them, "How long do you plan to be in town?"

Mr. Liang replies, "We are only staying today and tomorrow, then leaving the morning after tomorrow, and we have two more appointments before we leave, but so far, I am very impressed."

Neiman asks, "I have an idea of what you would like, can I get something to you before you leave?"

He responds, "You can have it that quickly? I am impressed, but we are looking for some very tightly locked down numbers. I know how long these things take to get worked out properly. Why don't you take your time, get very specific, then send it to me?"

Neiman states, "Oh I absolutely will, but I want to give you something to take with you to mull over, and send a formal proposal in a few days, and I would like you to meet the company president, Ben, before leaving. Can we meet with you here again tomorrow evening before you leave?" They agree.

When Neiman returns to the company, he reports to Ben that, "They were only staying in the city for the day and were off to other engagements elsewhere. I asked them to consider giving us a whirl, and mister Liang said they were there to expand their line through American vendors, but they were very skeptical of our size. However, they really liked the Consuelo line. I got a pretty good vibe, but the size issue was a real big concern. They said if time permitted, they would try to stop and meet again before departing the country."

Neiman failed to mention the fact that he had another appointment with them, or that he had invited them to meet the owner Ben, the next day.

The following day, Neiman leaves for the evening, and keeps his appointment with the 2 representatives at the restaurant in their hotel. He tells them, "I'm Sorry, Ben was not able to make the meeting, but sends his greetings, and he asked me to give you an overview of why you should do business with us."

After going through some initial marketing publicity fluff and details about the company, he gets to the meat of the matter, "Now that you've seen our operations, I have to let you in on a secret. Its closely held by the company for competitive reasons, and only disclosed to special clients. We were hesitant to show you until I was sure of your level of interest and the scope of work. What you saw was only the front operation of the company. On large orders, we sub-contract the type of product you want, out to a specialty subsidiary called NDS manufacturing, and doing business directly through the subsidiary will save you an additional 5%, and that makes us the lowest in the industry. The quality and workmanship is guaranteed the same or better going direct, and because we have the advantage of scalability, size is not an issue. An initial order the size you indicated will allow us to put you at the head with a dedicated staff, on a dedicated line, and I personally oversee the work done."

A few days later when Ben inquires about meeting the customer again, Neiman tells him, "Yea, I talked with Liang yesterday, and unfortunately, they were not able to meet with either of us again due to a scheduling conflict. Not to worry though, I was able to get him a worked up bid, which he said they would consider. He commented that he was impressed with the company's shop, and a few of the lines, but he really felt that our size was an issue for them at this time."

Neiman takes the opportunity to have a few of Ben's garment workers enhance the garment X-Packs™ which were purchased for the leased ARCH6, and already contained in the system, by demonstrating to the machines some of the secret trade 'tricks' innovated by Ben's father, which is some of what made the garments produced so desirable. The copied techniques were trade secrets and responsible for much of the exclusivity, and a great deal of the demand for their products.

Sun-Walk Fashions had asked Neiman for design workups and samples, so using the companies designers, Neiman produces them without the knowledge of Ben. He sends them the designs and a few samples and the proposal is accepted based on high quality and workmanship, and a 7% competitive advantage. They place a very large initial order for 5 of the possible 20 items in the line through Neiman's non existent NDS manufacturing, so he steals the business. The single order is more than Ben's company does in 9 months, and the indication is for another 6 to 8 times that order size over the next

24 months.

Neiman starts the Sun-Walk work order at the company, intending to use the internal staff and the ARCH6, but has trouble meeting the deadlines due to the workload. He attempts to shift the workload around at the company, favoring his own illicit business, however, he starts to fumble and is being contacted by a nervous Asian customer. He has numerous contacts in business and obtains a loan for leasing 2 ARCH6s and a single CHARLIE which were readily available and deliverable in a week. He recruits a few garment workers for support and rents an empty warehouse and sets up shop. He uses his accrued vacation and leave time, so he is missing considerable work at Ben's company as he is busy moonlighting to get the orders placed for the customer, who upon seeing the deliveries start to improve, is happy and payment flows readily. The ARCH6s' and the CHARLIE work non-stop 24 hours a day, 7 days a week, for 14 weeks to complete the job. They are able to complete the entire order 3 weeks faster than the original contract stipulated, so the customer immediately places an additional order.

The Asian retailer is so happy with the work, that the president sends a very complimentary letter to Ben expressing how much they are impressed by the work his company, it's subsidiary, NDS, and Neiman have done in quality and speed to produce the order so timely.

Ben is devastated. He respects Neiman and will not allow himself to believe what his gut already tells him. Not wanting his worst fear confirmed, he confronts him in a nervous but careful manner, "I got a letter from the president of Sunwalk, one of the largest Asian retailers who was here a few months back. You remember, you showed them around? They are very impressed with how we..., no wait..., how you, produced an order for them so quickly on a contract. A contract that I was not aware I had signed. Please explain the situation to me, because I am having a real hard time understanding what they are talking about?"

Neiman sits down and speaks with his brother inlaw in his office, "Well that is correct my friend, we do not have a contract with them, I, have a contract with them. When they were in town a few months back, I took the opportunity to present to them a company that could deliver to them what they wanted. They would have gone elsewhere because this company, your company, could not handle a job the size they needed, so I took advantage of the situation and put a deal together with my company."

Ben is shocked, and starting to move away from a state of fear while entering into a low but increasing level of agitation, "What is this my company, what company are you talking about, what is your company?"

Neiman smugly replies, "My company is NDS, for Neiman D Susskind, and my company can handle the jobs your company can not, so I take the overflow from here."

Ben, now getting red-faced, angrily demands, "You have no right to take anything from here without my knowledge or approval. What have you taken from here? I welcomed you into my family, I gave you a job, my sister, and most of all, I trusted you, and this is how you repay me? Tell me why I should not fire you right now?"

Neiman retorts, repeating himself in a loud and agitated voice, "You gave to me, you gave to me? Who do you think built this business? When I came here, you were already circling the bowl, and I pulled you out. You only managed to squander what your father built, and now you want to fire me? I have news for you my friend, if I had not accepted your offer of a job when you came begging me, this place would have sunk years ago. And now I have some more news for you; I am going to make you a partner in your own business, or I am going to buy you out, take your pick. If you don't believe me, just go and ask your sister, cause she owns half of your business, my half. Either way Benny boy, you don't have much of a choice. If I leave, you'll have very little left, and I'll just come and take what remains anyway. You can fight me on this, but you already know, you'll loose."

Having just hijacked the largest possible contract right out from under Ben's nose, and informing him that he will be left with nothing, Neiman now proposes a merger and 70/30 partnership, merging Ben's company with the now existent NDS manufacturing, giving Neiman the larger split, and leaving both Ben and his sister Chloe sharing the smaller split. Ben is not amused, nor very happy with the hand that he's been dealt, but he reluctantly accepts being bought out by the man that is an obvious gambler and risk taker, and plays a game of his own questionable ethics and behavior.

The Autonomous Economy I - The Strong Man

Neiman is an extraordinary man of risk and bravado and Ben knows this from first hand experience about him and still believes in him. Despite the ass-kicking, back door ethics, and double dealing nature of the man, he sees him gaining some pretty high ground going forward.

Neiman goes home and informs his wife Chloe of the good news and tells her, "You should have seen his face, the big baby, when I informed him that I was going to take his company from him. I told him his father Victor was ashamed of him, and that he told me so."

Chloe asks, "Daddy said that about Ben? I don't remember him ever feeling disappointed in his son. Daddy was always pretty proud of him, and thought he would do well, I thought. Well anyway, it is ours now."

Neiman hugs Chloe, and then asks, "Hun, did you get the big steaks that I asked you to get? Where are they, in the fridge?"

He opens the fridge door and finds 3 large beef steaks of about 2 pounds each on a plate. Chloe replies, "Yes I got 3 of them. I thought you wanted to celebrate and I figured you wanted one for each of us, you, me, and Ben, because he readily accepted the proposal."

He replies, "He did readily accept the proposal, and I do want to celebrate. Get yourself ready; tomorrow we are heading to Vegas..., to celebrate."

Chloe asks, "Vegas, again? But dear, you promised me..., no, you swore to me you were through."

He hugs her and kisses her and replies, "I did, but now I'm on a roll!"

He grabs the steaks out of the fridge, and walks out to the back yard where there are 3 large dogs in a pen. He walks over into the pen area, and the dogs go crazy barking excitedly while jumping up on him trying to get a sniff and taste of what he is holding. He grabs the first steak, and taunts the dogs by repeatedly holding it down just above their reach, then jerking it back up, until one of the hounds finally tears it away from him. He repeats the same process 2 more times, disposing of the other steaks with the other dogs, while Chloe watches dumbfounded from the window.

Prestonward

Prestonward Fabrication & Machine, a mid-sized manufacturing company in Chicago, Illinois, employing about 150 people, and specializing in products for the agriculture as well as the mineral exploration industries.   The company had placed an order with Dalbots for 2 of the ARCH 600s, about a year and a half earlier, and received them 9 months prior. The machines adapted fast, taking 2 months to be productive. It may be better to say that it took the company 2 months to adapt to the machines that were ready able to work in welding, sheet metal fabrication, and assembly with the installation of a few X-Packs™, being productive very quickly. They had proven their worth by reducing the labor cost of 2 of the company's top line of agricultural produce sorting, inspection, cleaning, and packaging equipment.   One line took 864 man-hours and 6 days to produce a single unit.   Prior to the arrival of the Dalbots units, it required a team of 18 workers including 3 welders and 5 fabricators to produce, but now took no welders, and only 2 fabricators, dropping the requirement to 576 man-hours, and a total labor cost reduction over one half.

The change and work reassignment occurred without incident or loud protest by employees due to the fact that the excess welders and fabricators had been reassigned to other lines the company was eager to expand into, and the fact that a machine is not technically a person, so union membership and rules do not apply, and the company vigorously defended that position.   At the last quarterly financial review and statement by the company's CFO, it became apparent that the bottom line for that particular product line had risen dramatically, and after obtaining another large line of credit from a bank all too eager to extend a loan to a company showing such dramatic increase in profit, the CEO placed an order for 2 more CHARLIEs' to be delivered within 6 months, and 3 more units to follow that.   All involved were sticking their necks out considerably, believing that both the company's and the bank's good fortunes would continue.

Trygve Kettilson, is a humorless, hard-nosed shop foreman, who comes out to crow and taunt the crew about a forthcoming announcement coming up at the weekly company meeting.   It is usually attended by all employees, both management and labor, and addressed by both the companies original founder and long time president, and the current CEO. He walks out into the middle of one of the production floors, and sounds a long blast on a horn in order to get the attention of the 15 or so crew, over the thunderous roar of machines and power tools. He squawks loudly in his distinctive heavy Swedish accent as several of the workers busy at their stations stop and shut down their machines, "Ok, listen to me all of you rats. I am only going to say this once more, so pour the wax out of your ears. There is a company meeting in the main conference room tomorrow at 1 PM sharp. You are required to attend. Don't be late."

One of the men, Parretz, mockingly says to him, "Trygve, say 'guggenheim' for me." Several of them laugh and snort at the remark.

He replies, "How about I say it this way, you're fired, get out!"

The man makes a mocking body gesture, and replies back, "Sorry Trygve, nice try, you can't fire me or any of us!"

He replies and gestures toward one of the ARCH6 units, "Not yet, but most of you rats are going to be fired soon enough. That is why the meeting is called. You see that robot over there, that is the future here, while your future is outside the door. They are going to announce your non-future, starting with you Parretz, and the rest of you rats will soon follow him out. But you know, I have to thank you for the rat sacrifice, it is because you are all on your way out, that my bonus is going to be so big this year. I am going to get a brand new Mercedes this year."

Another man, Perkins, says to him, "What makes you think they wont replace you first Trygve, what makes you so special beside having your nose up both boss's ass? And, why don't you join the 21st century and use CATs like the rest of the world?"

He says, "CATs are for rats, like all of you. Mercedes are for those that have a job, and have value for their employers. Enjoy the time you have left, rats!" He walks out of the room.

The meeting opens by announcement of several items of company business by the CEO, who then turns the floor over to the company president for a few remarks. The company's original founder, Terry Ward, says, "Thank you Frank, I just have a few things to tell you.

Mostly that we have had a record quarter in terms of profit, and you are all going to get a bigger bonus share this year. We are also going to be taking on a new line of equipment. We have acquired the Buckhead line, and will be fabricating it here starting in March. We are going to take building 2 out of mothball, starting this week to bring the new equipment online, so I am going to ask a few of you to put in some over-time to get things moving in that direction. That is really all there is for now. Any questions?"

Perkins asks, "Trygve said you are going to acquire more of the robots. Will they be put on the new line, or will the company get some new people, because there is no way we can handle a new line on top of the workload now?"

The CEO Frank says, "I'll take this, umm…, we have not decided on workload adjustments or assignments at this time, but as for whether there will be additions, I can tell you pretty confidently, that the improvement in profitability comes as a result of the addition of Dalbots machines, so we will be expanding in that direction, and not in human workforce. However, you can all rest assured that your jobs are secure for the indefinite future."

CHAPTER ELEVEN

Auto Fleet

Desktoping

Generally speaking, at any point in history, considering the state of the art for any technology, to some extent, overriding economic factors are what drive the direction of industries, and when the right conditions force change, it is often along a course toward consolidation in which these driving economic dictates result in significant standardization. History is rife with examples; from the consumer product industries such as household appliances, to automobiles, to industrial process; a model is established by one or more innovative producers, and from that time until the technology is obsoleted, all examples that follow have a form resembling the original models with small variations, and tend to converge on an efficient common form-factor and utility. This results in less profit from competition as margins shrink, and a less varied definition, but a more utilitarian, and more accessible manifestation emerges.

In the 20th century, this process accelerated significantly. To a large extent the acceleration occurs because of what came to be known as 'Desktoping.' Desktoping refers to a process of moving away from vertically integrated, capital intensive, heavy industry, toward cottage or boutique production of goods and services. The term came about when referring to a phenomena from the early days of small but relatively powerful computers becoming ubiquitous. When the cost of computing diminished sufficiently, it tended to proliferate throughout the developed world. Prior to this, even light design and manufacturing was quite capital intensive just as with heavy industries, requiring teams of engineers, designers, and draftsmen, and a large amount of office and manufacturing space. The design phase of a project usually passed through many hands and took months to produce, then took many more months to create a prototype, then it was extensively tested, then redesigned, all before it was ready for the consuming marketplace. When computing had become cheap enough, eventually what had previously taken a room full of designers and engineers months to create, could be done with 1 or 2 engineers on a desktop computer using specialized software in a few weeks.

The term, 'Desktoping', came about by metaphorically conflating the concepts of the desktop on which the computer is sitting, with the software metaphor for a desktop workspace in a virtual office environment, where work takes place. All were integrated into the software operated by the computing system, and came to represent the work environment, whether in real physical space or in software space. Eventually the same process of moving down the capital cost scale, combined with the power of networking through the advent of the cloud, extended to many other processes and methods in boutique and cottage industries. Desktop publishing, desktop manufacturing, desktop design, desktop engineering, desktop pharmaceuticals, desktop genetics, desktop medicine, desktop psycho-shaping, and so forth. The concept and term 'Desktoping' was extended and retained even after desktop computers became obsolete, with the desktop metaphor moving completely into the cloud.

Standardization

Computing, more and more, became an extension of the individual human mind and to some extent, the substance of identity. When the computing environment moved from the individual's hands to the cloud, the basics of terminal hardware, software and operating systems became 'standardized' and commoditized, while the emerging, or cutting edge remained 'value added', so was still unique and competitively marketed. Newly emerging innovation such as more powerful processors or storage technology and software services, etc, were still competitive, but had to be made to run on what emerged as, and became known as a 'standard operating system', while new and innovative forms of operating systems had to be adapted to run on the standard for hardware platforms.

Early in the transition, the competitive landscape and variety of system choices was too complex and cumbersome and beginning to become a significant burden on a computing consumer, becoming unproductive and wasting considerable time on different competing platforms and formats, attempting to adapt the exchange of data between disparate systems which lacked interoperability. This gave rise to adoption of standardized hardware and operating systems, with standardized interchangeable peripherals, ports, and networking, all the while the entire cloud computing environment was driving a defacto common overall standard, because the cloud became increasingly where the data and processing was homed, so it was forcing the issue with the diminishing terminal devices of hardware, software and operating systems, and their makers.

Consumers wanted the benefits of computing and the utility of the technology, but not have to maintain it; and the pretentious appeal toward 'brand snobbery' was becoming unattractive. It came to where the average consumer was less interested if another had a 'Pickle', a 'MicroStake', or 'Hem'droid' brand computing device, since they all did basically the same thing with little differences, but saddled the user with additional burden staying current, attempting to interchange useful information and skill between systems produced by competing producers at war against each other. The consumer was caught in the middle and wanted out. The cloud provided the eventual escape, and the standard system made it very cheap.

Industry Transition

At one time, screws, nails, nuts and bolts were advertised and distinguished by their unique utility and sex appeal, so they were sold competitively, but eventually these items became a commodity. With the automobile, eventually the drive toward competitive advantage of a unique offering based on style and individual appeal declined due to the economic advantage of standardization, leading to the commodity status of the fleet, while the driver was no longer the nut behind the wheel.

Historically in the automobile industry, when a new technology was introduced to consumers, it was forced to fit the existing standard base of automobile form that existed, or it did not do well in the marketplace. This forced the production of a cookie cutter industry that proved less than adequate for all tastes. The business of automobiles and their manufacture gradually transitioned through various stages, away from the problems associated with heavy industry to become a more diverse cottage and boutique conglomerate industry, composed of large and small companies, in large and small cities alike. Standardization drove the course for interchangeable modularity; using advanced composite materials; employing computers and sensors to control the mechanical surfaces and motion of the vehicle; forcing old mechanical systems technology, which had been pioneered nearly a century and a half earlier, to yield and quickly disappear.

In the early part of the automobile era, the business model was to propagate petroleum based, large and heavy internal combustion engine automobile culture throughout the society, to support very large heavy industry and geographically concentrated companies, However that model soon gave way to a disperse cottage industry in which sleek new aesthetically pleasing all-electrics and hybrid models, comprised of advanced materials and designs, such as using composite carbon fiber with exotic metal alloys, compose chassis, and coach, with electric motor drives on 1, 2, 3 or 4 and more wheels.

This type of vehicle and mode of manufacturing production caught on very rapidly, and strong. Every conceivable configuration on the road was made; from long-haul heavy trucks to scooter, trike, RV, and 'go-cart' and everything in between. Even the ancient unicycle re-dubbed the uni-gyro, was fit with a gyroscope for balance and stability and an electric motor control system. They were eventually seen buzzing everywhere short haul transport was necessary.

Navigation, steering, power and performance, safety, breaking, comfort, equipment and features, parking, energy systems, efficiency, utility, maintenance; all were re-imagined, redefined, re-engineered, and transformed. Style gave way to utility; lighter weight and energy efficiency became the standard and the goal, and a competitive new market emerging with many new players producing innovative new models and services, employing new cutting edge measures; all drove costs down significantly. New technology was applied to an old industry. Light manufacturing such as that employed in the production of office equipment, and industrial machinery, was deployed for manufacturing automobiles, but initially, much of it on a very small scale, while the technology of the old heavy industries pioneered in the 1920's, yielded then disappeared. The wide use of plastic and other exotic lightweight new engineered materials was employed in structural components, where before, heavy structural metal was used; plastic injection-molding of chassis components; composite carbon fiber; light weight structural foam; modular plastic snap-on panels, fixtures and assemblies for interior and exterior surfaces; all served to severely curtail weight without compromising safety.

Some were popular exotic models, single piece ultra-light composite carbon fiber units dubbed 'ion board' and 'charge board', which were self contained, OEM, all electric power-train, and control-by-wire platforms, which were wholesaled to boutique coach builders who fitted standard or custom coach interiors on top. They designed the styling and features of the interiors and built the cabs, then sold completed units to the public under various brands. This model of production preceded the first wave and is a throwback to the horse and buggy era when boutique coach builders made custom coaches built upon a standard chassis. The units were then sold through numerous national dealerships. In the more common models, the entire car, chassis, and coach is molded in one piece. Many of the high utility transport models weigh less than 1000 lbs, while they are able to comfortably cary up to 6 passengers and haul a ton and a half payload. Early on, models had rechargeable power sources composed of electric battery or hydrogen power cells. Units would swap-out a rechargeable battery or power cell module in a self-serve robotic changing station, located in various places throughout an area, such as older petroleum refueling stations. Within a short period, great breakthroughs in low energy nuclear reaction, previously dubbed 'cold-fusion' technology gave them sufficient power through on-board, self contained nuclear power generators.

The state of the art is called 'drive-by-wire', so for the most part, all are driverless or remotely droned, and controlled or monitored by remote stations, so they are able to navigate themselves by various means using various technology including fixed land-based and space-based geo-positioning systems. A nationwide smart traffic grid is employed to track virtually every vehicle in service at any one time. A passenger will be transported by way of a predetermined destination; or they only need verbally describe the location; give an address or geo-coordinates or by use of a com device to transmit the information to the vehicle.

Early in the century this technology was novel and started showing up in limited use in certain cities then spread to others. Many models were introduced by numerous producers including the old heavy industry manufacturers. Prices for all new units dropped dramatically, due to the new technology and sales were brisk, and the novelty was very attractive. Individuals and families enjoyed the freedom of mobility without the stress of navigating through stop and go traffic.

Safety was improved dramatically by the fact that navigation and control is handled in software and not by the unpredictable nature and flawed response and judgement of humans. The units employ systems like close proximity perimeter radar, and com links to communicate with and track all movement by other nearby vehicles, pedestrians, and so forth, so each knows exactly how its neighbor intends to move at all times, and will accommodate the other, thus reducing collisions to near zero. Imagine cars having courteous drivers. In intersections, where cross traffic collisions were traditionally at the highest risk, smart beacons are employed to communicate with each vehicle and act as traffic control systems, assigning and coordinating the movement of each vehicle as it passes through. They move in unison in a fluid motion mimicking the movement of swarms, like those of flying or swimming creatures that travel in groups. Death and injury rates for automobile accidents dropped to near zero mainly because the systems eventually attained flawless operation, and because of the ability for each vehicle to know and predict the movements of all others in its sphere of contact.

Autonomous automobiles that serve all the population are the most popular mode of transportation. Eventually, the industry consolidated when it became apparent the economic advantage that a driverless general purpose transportation vehicle posed, leading to the idea that ownership was less viable or desirable because it was cheaper by far to just order and employ utility transportation whenever it became necessary, so the industry was consolidated under a single umbrella in which many companies participate and control it all. It becomes somewhat of a public utility at this point even though there were still numerous private companies in diverse aspects of the industry.

CAT Fleet

For the most part, automobiles are no longer owned or kept by individuals, but fleets are run and maintained by numerous companies that cooperate and compete in a shared market pseudo-cartel system, locally regulated.

A few companies started buying large fleets and operating them as a taxi or transport service for those that are not able to afford an automobile, or have very light transportation requirements; elderly, and low income patrons were the initial target market niche. They could buy limited to unlimited transportation within a geographic area for a modest monthly fee. Early on, there are several manufacturers that made contributing entries including the legacy car companies, but they seemed to fall behind faster than their more contemporary newcomer competitors, or they were outright absorbed by the younger companies and not the more likely reverse scenario.

Individual ownership diminished rapidly due to the fact that with a monthly subscription on competing fleets, there was no longer any need to own an automobile, so the expense dropped considerably to about one quarter of the expense of automobile ownership. No need to pay for lease or purchase, fluctuating fuel price, insurance, maintenance and repair, garage housing, and commercial garage parking, and they became much safer due to the fact that the dangerous element, the idiot behind the wheel was removed. Now there were none of the worries associated with ownership, or the threat of death or serious injury by automobile. It was a no brainer, and despite the reluctance of mostly Americans to give up their personal autos, considering the alternative, the advantages vastly outweighed the less important loss of personal grip on the wheel and other forms of vanity, and the emotional enhancements associated with the ownership of metal. The fleet units were extremely prompt, safe, reliable, and cheap. Many people, after relieving themselves of the burden, converted their garages into more useful game rooms, bedrooms, or just left them as a dirt or work room. Massive tracts of land previously reserved for parking individually owned automobiles, were converted to other more useful facilities like parks and new buildings.

The business concept started in foreign locales, where several companies pioneered and offered the service. They were way ahead of the US in the use of economy transport services, so some US companies emulated the successful implementation. It started to catch on in the US when a company by the name of Reli-Trans offered a service called 'minute ride', and built operations in several large cities, then started buying up their competitors and employed a powerful nationwide advertising campaign to sell the idea to the consuming public; it caught fire and spread fast. Now a consumer of transportation had the choice between the 'pride of ownership', along with the significant financial obligation associated with owning and maintaining an automobile; lease or purchase; insurance; operator's license; fuel; maintenance; and paying for a place to house one's cherished object of such beauty and pride; or, simply buying a monthly pass that was less than a quarter of the average cost of ownership and maintenance combined for unlimited utility transportation. If they were in need of longer range transportation, the same companies gave them that option. American business being what it is, Reli-Trans acquired and consolidated the operations of several other services, creating a loose cartel of sorts, operating nationwide in a defacto monopoly, then renamed the holding company Consolidated Autonomous Transport, or CAT for short.

A CAT can be in your driveway every morning, noon, or night, or anytime, on standing order, or can be ordered to show up wherever you find yourself and arrive in 1 to 10 minutes. It can be summoned via a com device, or can be pre-programed to arrive wherever you are or plan to be whenever you need it, or meet you at a specific time and place. All the rider need do is order the unit in advance in the fashion of ordering a taxi, but through the cloud; or they can have one on standing order as in the old days of limousine service. The single rider, or group will climb in and inform the unit of the address or otherwise instruct it where to drive, or where to turn and stop, etc. They can plot the course in advance and transmit the route to the unit via various means. Upon being summoned, the unit will leave the idle state or storage facility, then arrive at the designated location just prior to the appointed time spotlessly clean, fully fueled, and ready to go. It will deposit the payload at the destination with exactly the same precision. Once service demand is finished, the units will go on to other assignments, or return for service, cleaning, repair, or storage, always out of sight, in super facilities that dot all cities, but still in close proximity so their arrival is very prompt.

They are always available at all times, on the street, commercial shopping area, grocery store, theatre, or the center of town. If, for what ever reason, summoning method options are not possible, and transport is desired, a simple loud acclamation of 'transport', or 'cat' will usually suffice for nearby or passing units. One can hail a CAT as easily as one could hail a taxi in a bygone era, simply by waving the unit over. It can be ordered to wait for you until you return up to an hour, or be instructed to return at a later time.

When not in use, they are housed in various locations distributed widely down to the neighborhood level, at very close proximity, but completely out of site. Some are in large vertical silos built upward or underground, moved in and out by elevators, where they can be compactly stacked vertically one atop another, and dispensed like diner plates. If out in the field, and in-need of servicing or renewal of energy, it will autonomously return to the home location, where it is serviced by the home system. Various models generally have the energy capacity for between 10 hours and a month or more of continuous operation before needing to be recharged. They are generally non polluting, and their energy systems vary depending on utility and make, and range from all electric or petroleum fueled, variously powered by low energy nuclear reaction, chemical or biological fuel cells, or are hybridized between several.

The cost is very affordable at less than a fifth of what it costs to own and maintain an automobile. Transportation is a service needed by virtually everyone, and the CATs serve that requirement very well. The immense economies of scale gained by the smart fleets, guarantee the most efficient means of paying for it, at any level.

Lunch

Circa 2046; Two young men, employees of a mid sized company in the heartland of America, one of which is white, and one Hispanic are sitting idly by and loafing in the office. A young black man, Jermal, walks over to where the 2 are sitting and seats himself. Its friday and they are engaging in small chat and banter as lunch time nears. They are trying to decide what to do for lunch, so they include Jermal. The decision seems to be narrowed between the 2 choices to either go to the company lunch room, or go outside the company for food.

Jim Humphries, a young man in his early 20's says, "I'm not sure I can face cafe swill again today. Who's up for crackle?"

The others get up and stand and stretch along with Jim, then head for the exit. They step right up and into one of several CATs lined up and waiting for customers just outside the building's main entry.

Once they had piled into the vehicle, Jim says, "To Crackle my good CAT."

The CAT immediately drives off with the boys toward the main artery and then onto the 12 lane highway for about 20 miles. The young men are in generally good spirits because it is Friday, lunch, and they are looking forward to the weekend. Jim likes to spend time on his hobby of restoring old manual driven automobiles, and he also visits his mom and sister on occasion. After the work day is done, Orlando is planning an early dinner surprise for his young girlfriend Dulce, and then going to a party at his aunts house. Jermal is quiet and does not discuss his life outside of work with his co-workers, so the others are not particularly interested and don't ask.

The CAT has 4 seats for carrying passengers arranged 2 on each end and each side. The vehicles have no conventional front or rear, but are able to drive in either direction. The unit is 6 foot across the exterior, with 5 feet interior breadth. Typical units have 4 doors plus a hatch for luggage, accessible from the inside and out. The seating inside of a CAT is arranged so that each seat may face toward the inside center where passengers may face each other. Each seat's position is controlled by a soft-joystick. It reclines and swivels on a keyed rotating platform, which is geared to ensure proper lateral placement for any given orientation, which allows the occupant to face in any direction up to 360 degree rotation.

The doors open automatically, and upon engaging one, to enter, each seat will extract outside the vehicle allowing the occupant the ability to climb directly into the seat, which then retracts back inside as the door closes itself. Each seat is self contained with a fixed foot rest that can be retracted out of the way. The seat also has a sliding panel that comes up vertically from the right side. It slides up and across the lap in an arc, giving a work surface and a cloud communication interface with cloud display monitor to accommodate the worker/commuter. Each occupant may ride and look wherever they desire, not being inclined toward any particular direction. The cloud-monitor will recognize the identity of the user/occupant, who can also optionally require a biometric identification, or personal password for security, whereupon by invocation, their personal cloud workspace is presented. Each seat knows, and automatically adjusts to each users particular preferences from the seat recline and position, to choice of entertainment such as music, and vantage.

The vehicle is completely autonomous in operation however, it can be manually navigated electronically from any of the seats, with the driver facing outside through the window or by looking at the com screen in the seat. Most models are also equipped with a small separated alcove which allows limited manual override of mechanical systems and is accessed from one of the seat positions. Behind each end seat, there is access to boot storage, with a rack for stowing personal items while in transit. In most models, depending on the pass level, up to 4 people can ride in a vehicle, but there are models that accommodate up to 8. In cases where there are large parties of people that desire to travel together, multiple CATs can be electronically 'trained' to closely follow one unit acting as lead as if they are physically attached, or for carrying extra luggage, cargo, or supplies. There are also autonomous drone units designed only for storage which may be linked to travel together in the same fashion.

As they travel to their destination, Orlando remarks, "I'm glad I was able to get a 6 month CAT pass because money has been a little tight." His mother occasionally needs to borrow money from her sister, his aunt, so Orlando also helps her out on occasion and recently needed to pay part of her rent, as did both of his sisters.

Jim replies to him, "Personally, I hate these things!", as he glances out the window looking at a sea of CATs with what Jim considers to be zombie occupants, "They are ugly and they all look the same, just like peoples houses, no individuality."

Orlando begins telling a story, "My grandfather owned a car for a while, so my dad grew up with the family car. He told me that they like the CATs now because it was hard to afford a car before then."

Jim asks, "What's hard about owning one, I own two?"

Orlando replies, "Yea, but do they even work? My family had to rely on their car everyday; there were no CATs back then. They had trouble paying for it; they had to get a loan from the bank and make payments, and banks don't exactly welcome mexicans. Then the bank made them carry insurance, which is very expensive. Then, before they were finished paying the loan off, the car was already worn out, so they had to buy another one while still owing the bank for the old one. They constantly worried about someone scratching it or crashing into it, and had to put fuel in themselves every time they wanted to go somewhere. Like I said, I am glad I have a CAT pass. Even if it's expensive, it's no where near as expensive as owning a car."

Jim replies, "Ok, dude you convinced me; I only own the ones I have because I like them, but I would not rely on them that much."

Orlando then asks, "Don't you have a garage where you store your cars, because my grandparents had a garage at their house. Now its mi abuelos office, and a game-room for los nietos."

He continues, "My dads brother borrowed the car once and got drunk, then he crashed it into another car. He was arrested because they said he was drunk while operating it, so he had to go to jail for a few months. Then, my parents had to take the car to have it repaired and had to rent another car. My uncle lost his car operators license, so he couldn't drive. I don't need a license, do you? Garl! Then even though he had insurance, the family of the car that was hit sued my grandfather and took a lot of money from the family. The more I think about it, what a pain in the ass it was for them to own and operate a car. I am really glad we have CATs."

Jim replies, "Ok, good point dude. I said you convinced me. A CAT is great for everyday use, but I still like my cars and I know they don't work yet but they will someday, and hopefully I will be able to pay for them then. Ok, we are getting close to crackle."

Jermal says, "My dad had a car for a while, garl, that car snitched him out. He went to visit my aunt at her house, when he was supposed to be at a game, or with his friends, and my mom found out. She thought he was cheating on her, so she called a detective, and he called up the car location on a map, and it said how he been spending a long time with her sister. Thats how she found out. Now they divorced."

Orlando remarks, "Garl, I don't want no car that will rat me out to my wife, or my boss, or the police. They also monitor everything you do when you drive it. If you're drunk, or speeding, or drive reckless, then your insurance goes up. Its just not worth it to have your car rat you out all the time, I don't want to own a rat."

Jim waxes poetically, and replies, "What about this rat, I mean CAT. Maybe it is a rat; you ever think about that? Don't you think this thing knows all as well? Don't you think this CAT will give up everything on you. Before you got in, it cloud-i-deed your cloud-ass, your cloud-self, and knows everything about your cloud-life. It can dump you out in the street if it don't like you, or refuse to even pick ya up. It knows everything we have said, and done, and where we have been, and going, and it knows the condition we are in. It knows if one, or all of us is drunk or stoned, or if one of us gets in or out, and where; if we have any alcohol to drink, or did something illegal. If you're wanted, it will lock you in and take you to jail. It recognized you and set your console and seat the way you like. It remembered that from the last time you got into a CAT, and set your seat and your console. You cant get away from it. Its the same everywhere you go, and everything you do. Everything about everything you do is known, although, I like what you said about not having to worry about the cost and headaches when you own one. Never thought about it, but it does sound like a pain."

Orlando concludes with the remark, "Yea, all that is true, but it don't matter about any of it. I don't really care what the CAT thinks about anything, and you don't have to worry about drivers license or drunken driving, or where you go, or where you been, or what you have been doing in a CAT. Its not gonna cost me any more money, and I don't have no secrets."

CHAPTER TWELVE

Cloud Self

The trace, remnant, or entity, manifest by identity, extension of mind and personality, imparted into cloud by presence, and manifest by activities; the mark left by activity in-cloud. The manifestation of presence with which an entity self identifies, and the identity with which others recognize the entity in-cloud. An avatar may be considered a cloud-entity, but not necessarily the inverse. One's cloud-self is not an avatar.

The term 'cloud' originates from the telecommunications industry in the US during the 1970s, when routing telephone traffic over digital communication networks first started. When analog voice signals were first digitized and packaged into electronic communications network data 'packets', they became 'routable.' Virtually all telecom services company's local offices were connected together through a network, with the capacity to carry services for local customers. If a customer wanted to have a dedicated private network or telecom link to carry voice or data between their remote offices, they could buy services such as Frame Relay, ATM (Asynchronous Transfer Mode), ISDN (Integrated Services Digital network) and others services that would connect them between remote locations whether separated by only a few miles on opposite sides of the city, the country, or the world, and without having to run a dedicated wire between remote locations. The data network operated over a series of local network switches and routers, which carried voice and data via a simple connection into what came to be referred to as the 'cloud', for long-haul transport, connecting local offices and customer voice and data together. The totality of connected data networks, taken as a whole is the cloud. The terminology was adapted, referring to the vast expansion of the same technology; the internet, and popularized.

The cloud now refers to a nameless impersonal collection of computers, servers, and storage that resides in internet space. The cloud collectively, contains a structure composing extension of the minds of all entities which have contributed and left their mark. It is the electronic cyber equivalent of the physical world and to a large extent, it seeks to emulate it. It can be characterized as a world that exists along side of the physical world in which we are born; a kind of parallel world. The data that characterizes much of what may be known about a person is stored somewhere, anywhere, or nowhere in particular, on whatever server is adapted to contain it.

Many large corporate entities built huge infrastructures of networks, servers, and storage systems designed to capture and retain data derived from various sources relevant to each and every person, wherever they reside, anywhere in the world, as far as they could be reached, whether engaged in the cloud or not. For these people, if not engaged, there were programs created and incentives designed to make sure they eventually would engage, by subsidizing the cost of access terminal equipment, given to them for little cost or free of charge. The objective was to get as large a portion of the worlds population as possible to engage in the cloud, then compile their digital profile.

Drive Thru

In the last 100 years or so, man began extending his mind and therefore his presence, outside of the limitation of his body by way of electronic computing facilities. Technology enhanced his memory, ability to perform calculation, gather information, organize his work, make decision, and communicate; things that in earlier times were severely limited, and came only by more primitive means. Prior to the era, he enhanced and extended his mind through tools and the written word, and before that, it was only through the spoken word. In all, he left an identifying mark. An axiom of science called the 'Heisenberg Uncertainty Principle' says that a phenomenon can not be observed without influencing it by the very action of observation.

The same principle applies to the cloud; one can not be a passive observer without influencing it by the mere act of observing, and more-so by engaging. Therefore, in the process of a man employing the tools used for extending and enhancing his mind, he has not done so without effect, but has influenced the world and left his mark. Whether intensional or not, virtually everyone has left a mark somewhere just by their presence in the world.

One's 'cloud-self' refers to the shadow, or signature, or distinguishing presence they have bestowed in 'cloud-space'. One's cloud self is not necessarily complete, nor possessive, nor voluntarily authorized by the entity to which it applies. It is as much a part of the essence, as DNA, or virtually anything else that may be considered identity.

§

On their way to lunch, the 3 friends experience the wonders of autonomous automotive transport combined with networking and integrated intelligence. They are riding in a CAT, which ably demonstrates the advanced state in the development of autonomous transport.

As they pass through the streets, it is obvious that they virtually never stop until they reach their destination, not for intersection, crossing, nothing. Even when pedestrians are in the street, the CATs gracefully maneuver around them. As the CAT which they are riding in passes intersections, it does not really slow much but passes all others from different directions completely unobstructed because an automated control beacon at each intersection sends out control instructions electronically to each CAT, and coordinates the movement of all traffic that approaches the intersection. The beacons are fit with Quantum Computing Devices as are the CATs' themselves, which gives both tremendous power to handle large matrices of data, to track, monitor, and control all the moving objects at once, as if each one were the only one in the field to worry about. It separately detects and communicates with each unit individually, and issues instructions to each as it proceeds, and all the other CATs detect and communicate with each other. Even in the absence of coordination from the traffic control beacon, they are able to work out and coordinate which car goes and when, all in milliseconds, and completely without the knowledge or intervention of the occupants who are in complete oblivion and unconcerned of the protocols operating the CATs.

The occupants are generally engrossed in leisurely self indulgent pursuits such as talking to others via telepresence on their console; indulging their 'cloud-life', playing immersive interactive virtual reality games; doing work from their desktops, known by their 'data-ID', or 'cloud-self'; talking with others in the CAT; or just staring out at the passing landscape.

Orlando is looking at his console, and pulls up a cloud-site, that he redirects to the main monitor screen in the CAT so the others can view what he is seeing. The group is still engrossed in the conversation about privacy, and he wants to make another point, "You guys see this sight? This is a group called 'Anon.' I heard about them, they teach people how to resist the all-seeing, and give you back a private life. You just have to learn to use anonymous credentials whenever you go in, or to just stay out. They say that officially, there are something like 20 to 40 percent more people that don't really exist because of all the anon fakes. That many people going totally anon is scary."

Jim Humphries retorts, "Those guys are not even updating that site any more. They're gone. There are no anon, not anymore. Most of them are in prison, last time I heard. First, they were busted for threatening national security, supposedly. You can't have privacy. Use all the fake names you want, they know who you are, even if you have 10 fake names, you're not anon, not anymore. Your face is your name."

Jermal says, "Garl, that so minionated thing to say, you dancin' on a string! My momma say never listen to a fool dancin' on a string. You a marionette?"

Jim replies, "I ain't yo marionette!"

"Garl, you somebody marionette!", Jermal retorts.

§

For a large measure, the data-self is fragmented and stored in bits and pieces all around the physical world. The entirety of your data-ID, or any fragment of it can be formed and reformed by calling together various segments related by some common keys which correlate the various bits and fragments which are spread out, and glues them together. Common keys refer to data that is common to the various sources of the fragments, which can correlate and tie them together. The use of common keys for the purpose of correlating and compiling all fragments into a comprehensive coherent homogenous record is called profile compilation. Common keys are past and present descriptors such as birth dates for self and all family relations, given name, nephonyms (cloud-names), age, spouse name, maiden name, name of children, employer, social or civic affiliations, street address, and so forth. They are varied, with some very useful and others not as much. The more longstanding, or permanent, and widespread amongst the fragments, and the more often the common key has been used to define or to identify, the more useful it is. In the US, the government issued number, which is often referred to as Socialist Serial Number, is the most common and therefore the most useful, because it has been in use for a very long time. It has always historically been utilized for the purpose of profile compilation and tracking the facts about, and activities of, the entity bearing the number, even before the advent of the cloud era. The number was designed for utilization throughout the lifespan and can not be easily changed or discarded by the whim of the entity assigned to it. Since the entity has been assigned to the number, it must possess them for the duration of their life. The cloud knows who every individual is, and virtually everything that can be known about them, and it can instantly recognize them wherever they happen to be.

§

As the 3 co-workers approach the Crackle depot, the fast food auto-teller starts to address each on his respective com screens. It can understand and talk to them in their own patois. As the boys approach, but while still about a quarter mile away, the cloud screens inside the CAT signal a message as the face of the animated, automated attendant, welcomes each of them. Crackle knows they are in-bound, and wants to have the order ready when they get there.

Each one of the 3 friends are greeted by name, and with a promotional offer, "Hello, how are you today? Can I interest you in our road warrior special?"

"No!", reply each of them as they stare at their respective com screens.

Jim is first, "Hello Jim, lovely day today sir, may we interest you in another number four which is what you had the last time on March 27th at 12:36?", asks the female movie star looking animated face on the screen.

"I've asked before, can you leave out the last time I ate here. I know when it was and I don't care enough for you to remind me again, ok! If I want to know what I had, I'll ask, and yes, I'll have a number four!", he replies, as he authorizes the transaction on his cloud screen.

"OK Jim, that will be one hundred sixty seven today on your account. Thank you for coming again.", the attendant speaks as it smiles back at him.

Meanwhile, the black face animated attendant facing Jermal asks him, "Jermal, Koo-anda qua-dol be in-da round-in wit do muh-fudin eat-nin?"

Which is to say, "Welcome Jermal sir, would you like to order a convenient meal?"

"Wit-in be do quad-in doom-anin, garls", replies Jermal, meaning, "Be so, but I am not getting out of the car, Garl!"

"Do be-ow-nin ki wid-al wit-in muh-fudin cred?", asks the attendant, meaning, "Will you be paying with gratuity or inconvenient credit?"

"Yo be mo-an-in an do-ob-in, OK!", barks an annoyed Jermal back at the attendant, meaning, "Don't be causing me no pain, OK!"

Jermal stares directly at the eye scanner.

"Som be no sha-ha-gin eva wit-in da niggas", he complains in a disgusted tone to his companions, which means, "Things are never ever really going to improve for the black man!"

Meanwhile Orlando is still staring at the screen looking and deciding what to order.

Orlando asks, "Muéstrame que una vez más..., hmm, no puedo decidir..., me mostrara la otra vez.", which is, "Show me that one again..., hmm, cant decide..., show me the other again."

The Hispanic faced animated attendant then says to him, "Lleve a su jef Orlando tiempo, quedan tres minutos", meaning, "Take your time Orlando hef, you are still three minutes away."

The car finally pulls up to the nondescript restaurant terminal and the machine attendant approaches and slides the orders in through the vehicles receiving port which the boys grab as the vehicle makes an exit.

"To Brammer's Park my good CAT", a jubilant Jim Humphries orders as the car leaves the establishment.

Cloud Mining

Prior to the cloud era, the cloud-self was considerably more fragmented and disorganized and variously stored and accessed from many sources of media, operated from various pieces of hardware, by varieties of software, but nonetheless, remained isolated and fragmented. The commercial expansion of cloud enhanced consolidation of all the pieces.

The information about an entity generally consists of whatever personal data is privately kept or shared, and some data controlled by the entity, such as work projects, documents, photos, audio and video content, personal contacts and correspondence, reference material, personal biography or resume, etc.; all that has accumulated in the cloud within the prior several decades; and, the dossier part of the information which the entity does not necessarily know about, nor has authorized, but is information gathered by other entities, such as accumulated financial records, tax records, police and criminal records, school records, insurance records, medical records, personal activities, interests, hobbies, proclivities, purchases, secrets, etc.; all of these can be characterized as accumulated records based on real activities. It may partially reside outside of the cloud in various forms, mostly written records not yet committed to digital form, but nonetheless a source from which the cloud may draw.

Then there are imputed inferences which are information about the entity that is guessed at through complex software algorithms, and based on the known information, attempting to fill in those pieces of information that are yet unknown. If the entity has interest in certain narrowly defined activities or products; is known to consume certain types of products, and live a particular lifestyle characterized by various descriptors, then transient and/or uncorrelated attributes such as age, sex, income, race, marital status, behavioral patterns, interests, and various other demographics, if unknown, may be inferred or predicted. These inferences can be bolstered by corroborating with other data.

The cloud-ID may be viewed as the external-ID, and may be characterized as an extension of the person, or ego, but residing in-cloud. Referred to by various adjectives; 'cloud-identity' or 'cloud-ID', 'cloud-self', 'cloud-life', or, 'data-life', 'data-self', or 'data-ID', to the extent that it exists, it is contained in the cloud and accessible from virtually anywhere in the world, wherever one is, through any device or channel, whether proprietary or public.

Labels are distinguished by what they describe. Cloud-identity refers to how others may know and recognize the entity through the cloud, and is distinguished by things known about the entity. Generally one or more nephonyms (cloud-names) other than given or birth names have been created along with a secret password to give exclusive private access to information and to engage. A cloud-ID may include distinguishing information about the entity such as the style of communicating, interests, or virtually anything that distinguishes one entity from all others.

'Cloud-Life' refers to activities in-cloud, and how an entity is perceived as manifest by those activities. It encompasses both recognized identity, and what is known about the entity by virtue of activities, just as real life encompasses activities engaged outside of the cloud, and is characterized by distinguishing known attributes.

Activities outside of the cloud become known and recorded in-cloud, and are attached to the appropriate cloud-ID dossier. With advent of all forms of digitally recorded data, from video derived via ubiquitous cameras, engaged daily, watching everything and everyone, everywhere, to the numerous 'eyes' of the ARCHIEs' and CHARLIEs', which record and archive everything they encounter daily, and given the fact that large numbers of them engage with humans all day, every day, there can be virtually no deed unknown. There is very little hidden from prying eyes, virtually every activity of persons is seen, and despite the fact no population ever gave either tacit or explicit approval of such a state, it has become entrenched, and pervasive.

Black Cloud

Arriving at Brammer's Park the boys finish lunch and set off to get back to work, but stop into the park's public restroom before heading back onto the road in the CAT. Orlando goes in, while Jim and Jermal stay outside. Entering, he stops to check some status at the cloud-com system at the wash station. The system is very similar to the cloud-com system inside the CATs. When standing in front of the wash station, the mirror also functions as a display screen, so the system will automatically identify the person through various means, then present a plethora of personally targeted ads and other information tailored specifically to them based on what is known about them, inorder to engage them in commerce. It will usually also offer their personal cloud-space desktop and the use of a telepresence session. This must be acknowledged and accepted via a token or other means of identification and authorization such as bio-ID, or facial recognition, or voice print and spoken pass-phrase. As long as the person is engaged, the system will continue to display their ID including a photo of them in a small section of the screen along with other information and tailored ads, for all present to see.

He finishes his cloud business and enters a toilet stall and is occupied, while outside an autonomous automobile arrives at the same location where a strange but very well dressed man goes inside. The man uses the facility and then goes to the wash station and engages in a conversation via the cloud-com system.

The man, oblivious to the presence of Orlando who is still in the stall, is engrossed in a conversation, speaking about an upcoming meeting in Wyoming. Orlando overhears him speaking to someone whom he calls by a very strange name.

Jim, sitting outside, decided that he also needs to use the facilities and went inside just as Orlando was exiting the stall. Orlando just stood for a moment watching the situation, then he makes his way to the adjacent wash station. The stranger in the room is still engrossed in his cloud-com conversation while pacing back and forth in the restroom, but when he observed Orlando exit the stall and standing there, and he saw Jim enter, he ended the session. He stared noticeably at Orlando, which made him a little uncomfortable, then nodded at both as they entered the area, then finished washing and grooming before exiting the restroom.

Both of them, having finished their business are at the wash station and Orlando says to Jim, "Urzababa? Who the hell is that? What kind of a name is Urzababa?"

Jim replies, "Yea I heard. Did you see the signet ring and the tattoos he had? Did you notice his screen; all black cloud stuff."

In the parking and boarding area, the strange exotic car is considerably larger and more elaborate than a CAT, so it is obviously privately owned. It stays for several minutes as the boys walk back outside and observe it is still present. They hurry and climb back into the CAT.

Upon entering the CAT the boys stare out at the car; Orlando says, "Did you see that guy? Did you notice his screen, no ID, see that is what I was talking about, he's anon!"

Jim announces, "Lets leave...., NOW! Take us back to work mister CAT."

Jermal, having observed the man as he entered his coach replies, "He ain't no anon, he minioned!"

Orlando replies, "Yea, I think you are right! He had no profile. It showed a black cloud on his screen where his face should have been. I heard about that, they can conceal their identity legally."

Jermal retorts with a little swagger and a subtle bit of humor, "You got the cash, you can get anything you want. He just buy anon. He protected; he minioned. You dudes were not supposed to see that." Jermal snickers in jest, "You ain't gonna live much longer now."

Orlando, a little frightened and taken aback squeals, "What, what are you talking about? What was I supposed to...., I, I didn't see anything."

Jim picks up on Jermal's joking manner and continues to chide a gullible Orlando, "Relax, he's giving you the business. Seriously though, I've heard about the black-cloud group as they are called. Supposedly they commissioned a company in Indonesia to design and build their networks and systems to monitor everyone and all affairs on the planet. They built and control it, and that's why we are in it and they aren't. They initially started it through the government, then took it over. The company in-turn recruited people from Indonesia, India..., uh..., and Asia.., if I recall..., then they had the architects, engineers, programmers, and builders all killed to keep it secret. We saw the guy's business and I'm sure he knows who we are and where we live... now I'm scared. Its like it was 40 years ago witnessing a mob killing. We were not supposed to see that guy in there, talking to someone!"

§

The minioned super-class naturally exempted themselves from scrutiny. Having always believed and appointed themselves as keepers, arbiters, and controllers of all speech, thought, and information, the necessity for the bogartation of all cloud-fruit was not likely to be overlooked by them. Although the fount of all things cloud sprang from many sources; the benign, to the truly inspired, and much with virtuous intent, the degenerate aspects employing the all-seeing, all-knowing, all-speaking, all-lying, all-intruding, agents, were unquestionably engineered by elitists, as a tool for the minioned super-class. They took measures to ensure they had access and control over the backend workings of cloud, to maintain and shape control; so they exempted themselves by way of filters, and a very comprehensive and profound black-protocol, to maintain their veil of secrecy. What had been intended for good, and considered a boon to mankind, and at the very worst, considered benign, was expeditiously co-opted by evil for it's purpose.

Co-opting implies control, and that is what transpired. Privately owned and operated cloud systems were subjected to legal requirements and regulations ostensibly for security, meaning governments were able to take total control of core systems. At will, they were able to shut down entire geographical regions of network segments, and route communications data anywhere. Intercepting communications streams for the purpose of intelligence was not the only function, it was also widely practiced to intercept and modify information sent to a recipient, en-route, inorder to distort or derail the message. Widespread spoofing; a method of masquerading as a known entity, used to implant false information with a recipient, was routinely used for manipulation. If they felt threatened by the activities of a single person or an entire organization, they would interrupt, modify the content, or shutdown the totality of message flow, wholesale. It was also common practice to financially ruin any entity that ran afoul of the brotherhood, through the same means.

Communications from any recipient became suspect, so confidence in the systems degraded steadily. Reality, consisting of accurate and factual information, imaging perception of the world, became a commodity controlled by the brotherhood. They had the power to completely shape the minds and attitudes of large segments of all nations. Like reality, history was written and rewritten to suit them. However, it was completely up to the individual, so it must be said that the power to control the perception of reality only existed if the recipient was willing to accept what was widely known to be tainted. It was a simple choice to reject all cloud-passed messages, whether news reports, educational, video, audio, text or otherwise, then subject the content to validation in order to correctly discern truth.

§

Arriving back at work much later than they should, they walk into the building, but as they are walking in, Orlando loudly announces in a rather distressed tone, "Oh, damn I left my case in the CAT. Man I cant loose that, I had a ring in there for my girlfriend; she is gonna kill me."

At that, Jim says, "Hold on, all you have to do is call that CAT and order a Lockdown, Secure and Return. Here let me show ya."

Jim grabs the com device from Orlando, and begins to show him how to do it. On the screen he pulls up the last activities section and it shows a map with the last route taken by the boys while in the CAT. The CAT itself is shown by a small icon which Jim clicks on, and the CAT's computer interface answers, Jim then orders the CAT to lockdown and secure, then immediately return, but he discovers that the CAT has subsequently already been occupied by other commuters.

Jim then says to Orlando, "Sorry man, but there is someone already in that CAT. Wow, I hope they don't notice your case, not likely, but they might just leave it."

Orlando says, "Not likely is right, it was not hidden, but just sitting on the rack."

Jim says, "You could call the CAT company dispatcher, but it won't do any good. I would just call the CAT and ask them to leave the stuff in the vehicle so that when it returns, you can get your stuff back."

"Good Idea, but why not call the CAT company dispatch?", he asks.

Jim replies, "They will just say they are not responsible, and only do what you can do yourself; put a Secure and Return order on the CAT, which you just did."

Orlando then says, "OK, lets call the CAT!"

Jim then clicks the call button icon on the screen for the CAT which then reveals a face inside the CAT who answers, "Ola, why are you calling?"

Jim hands the device back to Orlando who uses the controls to zoom out to reveal 3 menacing looking hispanic males in the CAT. Orlando says to the group, "Hi, I was in the CAT with my friends just before you got in, but I left something inside. I have ordered the CAT back as soon as you are done, so it will bring me my property. If you find something in there, please leave it for me to get back?"

One of the youths holds up Orlando's case and then says, "Is this what you lost homie? I like this, I think I'm going to keep it. Mi hermana Yolanda tiene un cumpleaños y quiere un como este, Gracias homie!", which in english means, "My sister Yolanda has a birthday and wants one like this, Thank you man!"

Orlando replies, "Por favor, hombre, deje, su para mi novia.", which translates in english to, "Please man, leave it. Its for my girlfriend."

At that he realizes that he is just making the situation worse by announcing there may be valuables inside the case. He glances over and notices that Jim is making a "button your lip" gesture, implying what Orlando already knows, that he gave away too much information, and now may have trouble getting his stuff back.

Orlando then pleads again, "Por favor hombre, yo realmente, realmente lo necesita. Por favor deje que para mí.", translating to, "Please man, I really, really need it. Please leave it for me."

The man replies, "Ahora me voy, hombre. Hasta luego.", translating to, "Now I'm going man. See you later."

He then reaches up and disconnects the com link.

Orlando is devastated and backs up leaning against the wall with a grimaced look on his face. He is distraught, but prayerfully speaks, "Dios mío, Dios me salvara. I lost that ring. I saved forever to buy it, now its gone. I was going to give it to her tonight when we go out."

Jim sympathetically and solemnly asks, "Is it an engagement ring? You gonna marry her?"

Orlando nervously laughs, and replies, "God no, its just a ring. Its a gift. God no! I didn't step in a trap!"

The Autonomous Economy I - The Strong Man

§

For those outside of the favored, virtually every knowable, discoverable bit of information is determined about virtually every living person on earth by gathering and analyzing the known correlated data, and examining it for discoverable threaded paths; chasing down the links in the data chain, leading to the virtually unlimited discovery and compilation of precision dossier. For those not protected, there is nothing hidden which cannot be seen. For those not covered by the black shield, truth and reality was a commodity to be had only by the most skilled at unravelling puzzles, or those able to construct and maintain covert private communication networks. The system and intelligent agents render everything discoverable and knowable to be discovered and known. With heightened efficacy, like the hydra which can not be resisted, relentlessly prodding, ever probing, threatening; it penetrates and seeks out the obscure; any space devoid of data. The information exposed, renders the entity vulnerable to exploitation, manipulation, and control.

Secrecy is like the truth of reality, and privacy; all are vital to normal human function, even for chattel. Like a pall, or blanket of confounded stupor across the land, falling like the ash from wide destruction; denial of privacy, like denial of truth, demeans the spirit, creating an emotional and mental state of imbalance, not easily reckoned in the mind, thus facilitating abhorrent behavior, cascading psychosis, neurosis, and paranoia; a highly volatile social dynamic, which must be adapted.

For those not initiated, tenacious inquisition, the autonomous intelligent agent, seeks the cold, stark, naked, secret; the inner sanctum of the heart, and mind, and is impersonal, and without gentility nor respect, nor the dignity afforded of decency. Shame is useful and common when exposing hidden deeds. For all but those protected, penetrating autocratic lying mouth, all-speaking, and the imperious all-seeing capstone eye, elevate guilt and shame as virtue; *'if your burden is contrition and shame, you are cleansed.'* No longer the need to demand; *'What have you hidden, or to hide?'*, all is known.

§

Entering their employers office, the boss of both Jim Humphries and Orlando walks into the lobby of the building where the 2 are discussing the matter. The man walks up to the 2 and while glancing at his watch he says in a petty, snarky tone, "Well its just..., oh..., about an hour past your lunch break..., so I'm sure glad you two decided we were worthy of you coming back to work. Orlando, you've been warned before. I'm sorry, but I gotta let you go."

At that Jim speaks up, and says, "Sir, we had a problem. When we went to lunch, someone stole some property from Orlando, and we were...."

The man cuts him off and interjects, "Well that sounds like Orlando's problem, what does that have to do with you? You're already on thin ice Mister Humphries. I suggest you not make the situation worse for yourself. Wrap up your business with Orlando and get back to work, while you still have a job!"

He then turns to Orlando and says, "I'm sorry to hear about your loss Orlando, I really am, but I have no choice on keeping you. You've just been lucky up to this point, considering. I have to let you go regardless."

Orlando, more worried about the loss of his property than the loss of his job, reaches out and shakes the hand of his boss, then replies, "I understand, you have to let me go."

The man turns away, and continues walking out of the building. Jim says, "Wow man, really bad day for you. Dude, I'm sorry again, wait, I know how you can still get your stuff back, call the cops and report it. You never know, they may get it for you."

Orlando replies, "Ok, nothing to loose now, lets try that."

He then pulls up the com device and calls the local police where he is greeted by the animated computer generated face of a stern looking officer that says, "Hello Orlando, what can we do for you?"

Orlando replies, "I just left my case with valuables in a CAT about 15 minutes ago, so I called the CAT to order a..., ah..., Jim what did you call that order for return?"

Jim replies, "Yes a secure and return order."

The officer asks, "An order for a secured return of the CAT?"

Orlando replies, "Yes thats it. Then I called the CAT, and others were in it, and they said they were going to keep my stuff."

The police officer face replies, "Please hold on Orlando!"

The complaint is then referred to another officer at the data center who pulls up the information on his terminal, which shows all the data including an aerial map of the area, which illustrates the route that the particular CAT took while Orlando and the others were in it. He pulls a video that shows Orlando and the others in the CAT, along side an aerial map showing the route they took, and where the CAT is now. The data screen shows the CAT is now unoccupied and headed to Orlando's location. The officer makes visual contact inside the CAT where the occupants, if any, do not know he is looking in, but it turns out the CAT is empty. He then looks at the recent route record on the map for where the CAT last delivered occupants. The record shows the location of where the CAT dropped the men. He then pulls up video of the men while they were in the CAT, and the software immediately identifies them by name, address, and their most recent location.

The officer then clicks on the icon for available CHARLIEs' on the street, and the system indicates there are 3, so he clicks on the icons for 2 of them shown by their location on the map of the area where the men were dropped by the CAT which engages both of them. They are both nearby where the men were most recently seen. He then sends an image of the man carrying the case belonging to Orlando to both of the CHARLIEs' and he gets back a few minutes of video imagery observed by one of the CHARLIEs' that witnessed the men as they walked on the street near where the CHARLIE was busy. He then dispatches a patrol with the complaint, the location, and the images of the suspects. The patrol quickly finds the men and detains them, and retrieves Orlando's case back.

The computer generated dispatch officer then appears on the screen of Orlando's com device and says, "Hello Orlando, we have tracked down your case, and dispatched an officer to secure it, however, I can not guarantee that we can retrieve it intact. Are you going to be at your current location for a while, where the officer can contact you?"

A very hopeful and relieved Orlando replies, "Yes, I will be here until I hear otherwise. Thank-you so much for finding it, thank-you!"

Orlando excitedly exclaims, "I think they found it!"

Jim interrupts, "Yea I heard, thats great man, that is great. Hey man, sorry about the job. At least maybe you will get your case back though, and you can always get another job, but that ring, well you may not be able to get back what that may have cost you, eh? Hey I've got to get back in there, while I still have a job. Maybe I'll see you at O'Rourkes on Saturday or even next Friday?"

Orlando replies, "Yea, I'll be there one of those days, hey maybe both, since now I don't have a job to worry about."

Jim extends his hand to Orlando, who grasps it, and says, "Thank you so much for your help. Lets stay in touch."

Jim replies, "That we will do my man. Stay loose, OK, and don't open the door to any spooky looking thugs!"

Jim then walks into the building. Orlando is waiting just inside the main entrance to the building and observes an autonomous police patrol unit pull up outside, so he walks out to greet the officer. The officer steps out of the vehicle, and hands Orlando the case, then asks him to sign an electronic slate to acknowledge the receipt. He opens the case and to his relief, his valuable ring is still inside.

§

As far back as history reveals, eventually light displaces darkness, and good prevails over evil. What was intended for good has been dominated by the powerful for evil. However, eventually that which gave advantage to one, turns causing those that co-opted it for evil to themselves be put to disadvantage. As always with everything comprising the forward progress of civilization, the advent of technology rises to serve the elite first, then progresses further to more fulfill its real purpose in service of the common, thus diminishing the temporal advantage of the powerful over the weak. Universal law dictates balance again be restored. For the continuity of civilization, of necessity, there must be a turning back through a process where equilibrium is reached; balance restored. In this context, it follows that the temporal advantage of strong over weak, which technology bestows, abates and reverses when the advantage gained disperses, and is adopted broadly.

The belief of certain primitive tribes, laughed at and ridiculed by the educated and sophisticated as superstitious and ignorant, because they refuse their photograph taken fearing it steals their souls, must now be reexamined in light of the new reality. Maybe they knew something we had yet to learn. The impact of the loss of privacy is tremendous, with moral and philosophical questions on the intentions of the minioned super-class; wars, and their plans for war, permeating corruption and criminality; all dependent on the cloak of secrecy. Eventually, there is no privacy, and that would seem to favor the powerful at the expense of the common, but that pendulum only swings so far, then rebounds giving tremendous advantage to the common masses. How do the objectives of a powerful secret society remain secret when everything is linked and can be known? Can the integrity of the black veil remain? Will nothing penetrate, ever? Can their intentions and deeds remain cloaked, or will they eventually be laid bare?

How will future generations view total loss of privacy? With their lives and every aspect of it, captured and data reduced, will we heal once again when the power to harvest the human spirit diminishes. With balance restored, will the loss of privacy be as threatening then?

CHAPTER THIRTEEN
Survival

Banished

Adam and Woman find themselves in a terrible state compared to previous. Rebellion complete, their world has transformed from a paradise without struggle, to the demands of nature with much labor against decay. Bramble and Thorns fill the fields where there were flowers and fruit; the sunlight fades and they are cold. Their bellies are empty and the fruit they gather rots in the basket. The worm and the fly consume their portion. Lord God has decreed mortality upon their struggle; misery and decay consumes daily. Entropy engulfs their world; they are bound without escape.

...And Adam called his wife's name Eve; because she was the mother of all living...

...Unto Adam also and to his wife did the Lord God make coats of skins, and clothed them...

Adam and Eve observe the Lord God slaughter large beasts, making their skins into clothes for them, giving context to their lives with first knowledge outside the garden, so they may be preserved from cold. Lord God withdraws from them, and they implore him, but he answers not; then they see him no more.

Season to season, they deliberate on their state and seek wisdom where it may come.

They pray the sun, "You shine your warmth on the trees and grass and love them to bring fruit that we may be sustained, why does your light fade?"

They pray the stones, "You give shelter to the beasts, where may we obtain refuge from the cold?"

They pray the earth and the stream, "You give life and growth to the trees and flowers, and the herbs of the field, where may we gather?"

They pray the trees, "You bring forth fruit that we may be sustained, why does your leaf fail and your fruit fade, when may we gather?"

Neither sun, nor stones, nor earth, nor tree answer them, but to say, "Lessen your poverty from our bounty as you may, and learn wisdom!"

In the passing of time, Adam and Eve adapt their new paradigm, how much a pale shadow of their former. Adam is occupied gathering seed on a hill and sits down to rest. He looks to the open field beyond a wide stream, offering a greater measure of the bounty he gathers. He wades into the flow attempting to cross the stream, but the swift current turns him back.

Eve has gathered grass, and woven it together. She has made baskets and mats for sorting and separating the seed of the grass from the shaft. She crafts lengths of woven grass and bundles them to great length and girth. Eve presents the rope to Adam. Adam gathers sticks and ties them together forming a raft. He fashions an anchor from stones tied with wood. He casts the anchor and rope opposite of the large and fast stream, until it is fixed, then fixes it on his side. He pulls the rope to taxi himself and the raft to the other side of the stream.

Adam gathers his bounty and presents the parcel to his wife, announcing, "Eve, I have overcome the barrier of the stream. I have trodden upon the far field, and have gathered; the seed is good. Tomorrow, I shall take rope and devise to climb the rocks high up, that we may shelter there!"

He has overcome the obstacle; he has triumphed over the demands and constraints of order; he has learned cunning, and wisdom to devise his advantage, to hold back decay. He and his shall live!

Adam sits with Eve in the evening meditating on their lives. He reflects on their life before disobedience. He remembers he was as a child in innocence and freely walked with the Lord God in the cool of the evening of the garden; he only need tend the garden. He remembers he knew not loneliness because the presence of the Lord

God was his contentment. Lord God gave him a helpmate, Eve, whom he loves. He remembers he was not acquainted with death, nor sorrow, nor hunger, nor decay; only plenty and warmth, and the pure love of the Lord God who was his light. Adam tells Eve, and they contemplate Lord God was indeed right to decree death and decay upon them. They seek redemption, and repentance of Lord God, and restoration to the garden.

Adam says to Eve, "When light comes, let us go to the midst of the garden, where abides the tree of life, of which Lord God has bade us eat. We have obtained wisdom to overcome; we shall obtain life forever. May Lord God restore us!"

...And the Lord God said, Behold, the man is become as one of us, to know good and evil: and now, lest he put forth his hand, and take also of the tree of life, and eat, and live for ever:

Therefore the Lord God sent him forth from the garden of Eden, to till the ground from whence he was taken.

So he drove out the man; and he placed at the east of the garden of Eden Cherubims, and a flaming sword which turned every way, to keep the way of the tree of life.

Life forever can not reconcile with death and decay so cannot be had by Adam, nor his progeny; it is a paradox. His is the domain of death and not life. He and his shall remain in death until their days are complete. There is no restoration to life and paradise. He and his spawn, from hence, shall learn to sustain and overcome through pain, hunger, poverty, and deprivation. Through violence and treachery, they practice and hone the skill that the knowledge of evil teaches. They multiply and overcome; they organize and build; they conquer and are conquered; each building upon the ruins of his brother. They construct cities and build great kingdoms, establish tribes and nations, and devise their own advantage, all the while seeking restoration to the garden of paradise.

The Organizing Economic Directive

[MD Narrative]: Since the days that humans lived in caves till recent, there has always been the need to organize ones self and ones tribe, family, community or

social order around one or more economic constructs. Most of human history is recorded by describing the events around an economic construct of one sort or another. The historical record always revolves around the Royal House, the nation, the family, the tribe, the military, etc. All are examples of things that evolve because of an economic organizing principal.

§

The most primitive, innate, most instinctual compulsion that has driven humans from the beginning of civilization and before, is survival, and the drive that compels survival is a prime directive of economic organization. Survival has always been elusive, and the drive for self preservation is the prime directive of survival.

Early on, humans organized themselves in small groups of tribes, and hunted, traveled, lived in caves together, etc. All are driven by survival. The driving economic principle that characterizes group-organization is that it is far more efficient for the production of food, labor, common defense, education, childcare, industry and so on, than the same for those not organized in groups. Living in herds under virtually any organization is a construct that is advantageous for the herd, and even animals follow the same behavior. It is the way nature has determined, and the basis for all socio-economic structure in civilization, from the dawn of man to the modern era.

Togtu and Ula

Circa 3500 BC, North-eastern Anatolia (Modern Turkey), a small band of primitive men and women are moving along a well travelled path down a slope that ultimately leads to the southern shore of the Black Sea. It is mid morning and Togtu (Toght-too) and Ula along with 4 others are scavenging for food, while 3 forward scouts with the party are hunting some small game. They are heading for a larger settlement of their people and just hoping not to run into the foreign band of others that have recently invaded the area. The other tribes are attempting to establish territory for hunting and foraging in this region because the land they previously inhabited has been suffering shortages of game and raids by others that have killed several of their tribe including their chief. Their scouting parties had recently been spotted coming in their small boats made from whole tree trunks, and tethered rafts landing in the area apparently coming from the North-

Eastern part of the Black Sea traveling along the coast while occasionally stopping along the shore to scout the region for resources, and occasionally running into the local tribes.   The invading tribesmen have arrived in a sizable group of about 8.   They are organized and skilled in quick hit and run raids, where their leader has instructed most of them on the finer points.

Strength in Numbers

[MD Narrative]: Cooperation in groups produces many benefits that the individual could not realize by working alone, allegorically based on the idea that one stick may be broken, but several sticks stacked together composes tremendous strength, and can not be broken. It can be summarized and named by the concept of synergy in human social structure in terms of strength and effect, 'The whole is greater than the sum of its parts.' The concept may be considered instinctual in that people will naturally organize themselves, and cooperate in numbers for the mutual benefit derived, without having first explored the relative advantages of such ordering.

§

Togtu and Ula are a couple so Togtu is very protective of Ula. He along with her, are leading their small band back to the small settlement to present the bounty of game and gathered plant and mineral materials from their latest expedition. Most of it is contained in woven bags slung over a large pole which is being carried between 2 men in the party and on the backs of a few of the women in baskets. Suddenly, out of nowhere one of the young men carrying the pole is struck in the leg by an arrow and drops down on the trail. He struggles to get up onto his other leg and scurries into the tall grass. The other pole-man also dumps his end and scurries off taking cover. The rest of the closely positioned group then immediately drop their cargo, and hunker down behind rocks and trees taking cover, attempting to see where the attackers are coming from.   3 of the women carrying bags of gathered food materials have dropped their loads and are hiding in the brush about 30 to 40 yards behind the lead males in the party. The attackers then make their way onto the trail and grab 6 of the baskets of food and run back into the grass and on into the woods hooting in glee as they flee. Togtu then runs back and helps the women including Ula who is still hiding.

In the encounter, the boy that is shot, Bartu (Boch-too), sees the attacker eye to eye and is in great fear for his life because the man pulls out another arrow, and begins to draw his bow to finish the job, when the leader of the expedition, Togtu, comes running back down towards the scene. The attacker then drops what he is doing and hightails it back into the grass with the rest of the band carrying off their load of booty.

One of the younger members says to Togtu, "Lets not delay, leave them, those bandits are getting away Togtu, and they have shot Bartu."

Togtu replies, "Don't worry about them, they left running toward Uctula, and the enclave, they'll probably run into Goyme who will capture them."

The Division of Labor

[MD Narrative]: If a man were to try to do everything by himself; if he were to attempt to manufacture all of the items his family consumes daily and perform all of his families needed services by himself, and he were to attempt to protect his family from criminals and the military power of others, by himself, twenty four by seven, he would not get very far. So trade and industry, defense, and so on, require cooperation and organization. Modern economic theory dictates that every person be willing to do his job in the social construct so that he can help contribute, but ultimately he acts in his self interest for payment of his share, for his effort, ensuring survival another day.

§

Togtu's band are working together to help establish themselves in the hierarchy of the tribe. Several are originally from a neighboring tribe that are known to Ula's home tribe. Togtu was originally from one of the neighboring tribes with whom, Ula's tribe had established a trade relationship. Togtu has received some respect and notoriety amongst the young men of the tribe because of his demonstrated skill, and because he is busy teaching others in the tribe the techniques and skill with weapons that have made him a successful hunter. In the short few seasons he has been a member of the tribe, he has made himself quite indispensable to them.

He has built a house in the compound of the enclave, where he intended to move Ula after making her his bride, which he did as soon as he was made a lieutenant under Goyme (Gho-ee-may), who respects him and treats him well. Ula is skilled with her hands. She makes decorative pieces from some of the materials she collects, as well as many other objects such as cooking utensils, and she also weaves baskets and cord from the material. She trades them amongst the members of the tribe, and if there are any surplus, she sells them to the merchant traders that come on occasion with things to trade. She is the daughter of the former tribal chief, the now deceased elder brother of Uctula (Ooghct-shula), the current chief, who took her in and raised her after his brother was killed.

Trade

[MD Narrative]: Just as each man, can not be self sufficient in producing all he needs to live, and despite the effort of the family's or tribe's ability to be self sufficient within itself, there are many things that may be necessary, or may only be considered a luxury, which nevertheless, can not be supplied by the tribe, and so must be supplied by others outside of the tribe. Trade relationships were traditionally developed through trust and confidence in the trading partner. Trade adds depth and dimension to a society, and is likely to increase the general wealth and welfare of the members. In some instances, the tribe was positioned to control trade routes, extracting a toll on each and every trader as they passed through. The trade system could greatly increase the wealth of a few that had the inside track on the rare goods that were sought after by the wealthier members of the tribe willing to pay a considerable sum to obtain them.

§

Togtu was invited to become a member of the tribe when he met up with the chief's son Goyme, and demonstrated how he could be of some worth to the tribe by hunting and bringing in large game. His move to the other tribe was partly motivated by getting close to Ula, whom he had met on a few of his trading visits. He was exceptionally skilled with a bow and arrow and a long hand held shaft with a hook on the end, referred to by the tribe as a Tetai, and elsewhere referred to as an Atlatl among other names, used for throwing medium to long shafted sharpened darts such as spears. In the hands of a skilled hunter, they are deadly.

Although Togtu was brought up in a neighboring tribe, he was often in the area bringing in game he had killed, to trade with Ula's tribe, when he met Goyme, who invited him to meet his father Uctula, the tribal chief. Uctula is an old chief and like Togtu, he was a skilled hunter in his younger days. He had given his blessing to Togtu as a probationary member of the tribe, which meant he could come and go, and was trusted by other members. After one season had passed, Togtu was adopted, so his probationary status had changed to permanent member of the tribe.

Economies of Scale

[MD Narrative]: Survival is paramount, but how well one, and by extension, one's tribe survives and prospers, ultimately depends on many factors which include how well economies of scale are employed. Modern mass production was pioneered by Henry Ford during the American phase of the modern industrial age.

§

The basic principle of economies of scale is simple; the more of a product or service that can be produced at one time in a batch or process, the lower the cost of each produced unit; because of the costs associated with the inefficiency of having to startup and shutdown production, and the cost of tooling. Tooling is very expensive, but larger tools are more efficient, so reduce the cost per unit, which more than pays for itself, given a large enough quantity of production. If a producer can simply start a production line, and keep it going forever, the production cost for each unit is reduced, and continues declining to optimum.

Also involved in this, is the efficiencies that come from specialization at each step in the production. A skilled worker that only has a simple job to do repetitively is more efficient than one that must perform several disparate duties. Combine this with the reduced cost of mass produced input materials, and optimum production is achieved. The cheaper the production costs, the more that can be made and the more available the product becomes, the lower the cost of acquisition, meaning the lower the affordability threshold.

§

Henry ford, the Michigan based founder of the Ford Motor Company, struggled with competition and was somewhat obsessed with maintaining what he believed to be first-mover position when it came to making and selling automobiles. He innovated around the idea of treating his workers well enough to keep them happy, and paying them enough to be able to afford what his vision entailed; that the automobile would become something possessed by the common person, and not just a curiosity or indulgence for the rich. For this to be accomplished, he needed to out-perform his competitors, so he had to come up with ways to produce more automobiles per worker, and discovered workers were more productive if they were happy, and if they were allowed to master doing one task well, rather than 3 or 4 tasks at less than optimum.

In addition, he realized the efficiency gains of having some workers supply parts and materials to the other workers busy assembling or machining parts. This he did by producing a moving production line so that each unit moved to several work stations successively, where the needed assembly tasks were attended by dedicated and very skilled workers. They were free to concentrate on doing what they were the most competent at, rather than expending time and energy moving about unproductively fetching supplies, or components, or bumping and stumbling over each other as each attempted to do their assigned task. With the older methods, things were not necessarily done in correct order, another level of inefficiency. The problem was solved by the line moving from station to station, so assembly proceeded in the exact proper order. Thus he innovated the mass production assembly line for automobile production and beat his competition at producing a larger number of cars, with higher quality, and lower cost. His competitors then emulated his methods and the modern mode of production was established.

As Ford was able to produce more automobiles per worker, even though they were paid higher than their counter-parts working for competitors, he was able to supply significantly more than his competition, and at considerably lesser cost to the consumer. As a result, Ford became the world's preeminent automobile producer.

Togo Africa

Landing at the port in Kpeme, Togo West Africa, one is struck by the abject poverty within 1 block of the port area. It is everywhere; It penetrates; it is profound… the depths of depravation are staggering. Everywhere one looks, there are groups of very heartbreaking people. The market bustles, but there are many children that are obviously without parents, begging in the streets. A little further down toward the south end of the city, there is the city garbage dump, with thousands of people sitting waiting for each truck as it pulls forward to dump its load. Most of the small orphaned children must compete with the larger more aggressive children and the adults that fight and beat each other for anything of value...

Pearls

Berry Gonzales, a reporter for the local network affiliate Cloudcaster - Philly-6 is reporting live and speaking, "We have a special story for you tonight here in downtown Philly."

A philanthropist that has a tremendous fortune, and lives in Philadelphia Pennsylvania, the city befitting the moniker of Brotherly Love and Philanthropy. We found out about him because a news crew has discovered the story and likes to give occasional updates because of the uplifting and heart warming sentiments despite usually following an otherwise shallow and vapid stories in a profession full of plastic Ken and Barby stuffed shirts, regurgitating otherwise inconsequential brain dead fluff. This is the second news report on the event, which was first publicized in November of 2046 with story updates now and again for the next 15 years, when it occasionally pops up and is given attention, usually every 3 or so years apart, mostly around the holidays. It comes into view from the background of everyday life because of the sensational nature of the story, and because people love to follow this type of story. First viewed on Cloud-TV, it grabs the attention of virtually everyone present. The philanthropist is shown giving away sizable quantities of money to cover the immediate desperate situation of 3 needy families, who spew forth with blubbering gratitude and unrelenting bear hugs, swelling the hearts of all, and making the ladies wipe their eyes with hanky and sleeve.

Reporter Gonzales is standing on the street corner with the generous gentleman, a microphone poked in his face asking him questions as an eager crowd of low income and poor, that usually line up for hand-outs, is clamoring to get to him first. A couple of policemen are seen in the background standing in front of a tape barrier used to keep order in the crowd.

The man's lawyer, or possibly his financial advisor, is also in the background with a megaphone instructing the crowd, "You must maintain a single line starting right in front of where I am standing, behind the yellow tape. Mister Kennedy will speak with all of you to hear your need, but you must remain calm and wait your turn. You will all be heard. Mister Kennedy has no desire to see any of you forgotten, please remain orderly, you will each get a turn to speak with Mister Kennedy."

Reporter Gonzales asks, "Can you tell us your name sir, and what you are doing out here?"

The man replies ,"I'm Julian Kennedy junior, and I am here to help the needy people of Philly as much as I can. Father Batista teaches that giving is good for the soul. These people need money, and I am in a position to help as many of them as possible, so I would like to hear from each of them what they need, uh, in an orderly manner of course..., then I will decide who and how much I can help."

The reporter asks, "That is truly a blessed thing you are doing. How much do you intend to give away, have you done this before?"

Mr. Kennedy replies, "Well I can not tell you how much I can give now. I started a few years ago. I made an announcement through the media then, but I guess most people thought it was a hoax or something because I only got a very few that bothered to take me up on the offer."

The reporter then asks, "I guess we are going to meet the first recipients here today, is that right? How many did you offer to help, and what have you given them? Oh here they are now!"

The reporter spins around to talk to the first of another group of 7 people standing in the background. A small frail black woman about 75 years of age is standing right next to Mr. Kennedy and the reporter asks her questions, "Hello ma'am, what is your name, and can you tell us your story, and how Mister Kennedy has helped you?"

The woman replies, "Well my house was on fire, and I had to go live with my daughter for the last 3 years, and she is real sick and not able to take care of me and herself and her two childrens, cause she is real sick and not well very much."

The reporter asks, "...and what is your name?" She replies, "Willa-Mae Beacon!"

The reporter then repeats again "…and how did Mr. Kennedy help you?"

The woman replies "He help me buyin a new house and help my daughter with being sick and help her doctor make her well, and pay for the hospital."

The woman's daughter is a large black woman in her early 50s and she is so grateful that she grabs Mr. Kennedy with a bear hug and holds him tight for almost 30 seconds, crying and wiping her tears away and saying thank-you, thank-you, thank-you Jesus over and over repeatedly. The reporter then pushes the microphone into her face and the camera zooms in to get a good close up, actually zooming in and crassly following a tear as it rolls down the woman's face.

Mr. Kennedy, with a tear in his eye, and choking back his emotions says, "Thats what it is all about, right here!"

Later interviews with others of the recipients, reveal that Julian Kennedy financed the desperate need for life saving operations, and replacement of homes and jobs after recent tragedy.

CHAPTER FOURTEEN

The Strong Man

The story of the human race is the advent of civilization which has been a very consistent story from the beginning of recorded time, and certainly before that.   That story and the history has been characterized by a struggle to obtain the substance of the support of life against the elements and the cruel demands of nature. From the beginning, when man first gazed upon the broad plane, and the incredible bounty that was God's gift to him, he began to wonder at the creatures that co-inhabit the earth, the others of his kind, and his place in the universe, and he realized that there was a contest for top position; to be the first; the top dog; to be *The Strong Man*, and that to lose may mean the end of his dream and all he holds dear. Losing means he is lost; second or third, and not as worthy to those with whom he held out the promise of a champion.

He is driven by his physical needs and the physical needs of those that depend on him, who have pledged their lives and devotion to him; his regal subjects, his ministers, his family, his children, his parents, his compatriots in the struggle, his world; those he defeated in the struggle, who now respect and admire, and follow him.   He represents their survival as well, and not just their ability to feed themselves, but also their ability to have warmth, and shelter, and pleasant surroundings, and the pleasures of life and comfort, and love and admiration, and the knowledge that you have lived a good life and survived as well as anyone could have expected.   He, and now they, have gone up against the best that mother nature could throw at them, and have conquered. This in contrast to the others who did not overcome, but lost and are no longer with us; they did not feed themselves; they did not have shelter, or warmth, or friends, or the respect of others that men need; they did not conquer, and they are not remembered.

Cain

...Adam knew Eve his wife; and she conceived, and bare Cain, and said, I have gotten a man from the Lord.

And she again bare his brother Abel. And Abel was a keeper of sheep, but Cain was a tiller of the ground.

Cain and his brother Abel were blessed by the Lord God, and so prospered in his sight. Of all in the tribe of Adam, Cain was a mighty man, and the most favored. He planted many fields with corn, and builded cities, and commanded many in the way of advantage to overcome the fly and the consuming worm. It came to pass that Cain was a mighty counselor to the tribe of Adam, and he became as a king unto the nation. He was uplifted in the sight of his brother Abel, and Adam.

Giving heed to Cain's strength, the tribe of Adam overcame much tribulation, and the people rejoiced in him, so Cain magnified himself in the sight of the tribe and became arrogant in the sight of the Lord God.

And in process of time it came to pass, that Cain brought of the fruit of the ground an offering unto the Lord.

And Abel, he also brought of the firstlings of his flock and of the fat thereof. And the Lord had respect unto Abel and to his offering:

But unto Cain and to his offering he had not respect. And Cain was very wroth, and his countenance fell.

It came to pass that the cities which Cain builded fell to ruin, and the crops of the field were lost; then the favor of Cain fell. Cain laid taxes on them, and the tribe of Adam grumbled because they were wroth to bear the tax of Cain. Their house had fallen and their corn failed; they had not substance. Cain raised his hand to obtain substance among the tribe.

And the Lord said unto Cain, Why art thou wroth? and why is thy countenance fallen? If thou doest well, shalt thou not be accepted? and if thou doest not well, sin lieth at the door...

Abel harkened unto the voice of the Lord God. He admired Cain and spoke praise of his brother. He was humbled in the sight of the Lord God, and did not magnify himself in the sight of the tribe, nor was he lifted up by the nation, because he was a keeper of the flock who followed the earth. Abel was blessed by the Lord God, and with a great flock fattened from the grass of the field, and many as to cover the mountain.

The tribe of Adam came to Abel because their houses had fallen and the corn failed; and the earth would not yield them the fruit thereof. They did plead with Abel to give meat to sustain their children and not perish, for Cain has taken their substance. Then gave Abel to them of his flock, to take thereof and eat.

Cain brought ruin to them and he did obtain their substance; so the tribe of Adam rejected Cain and uplifted Abel. They rejoiced saying, "As with Abel, we must follow the earth and tend the flock. Abel is greater than Cain!"

Now Cain was enraged and moved to jealousy.

And Cain talked with Abel his brother: and it came to pass, when they were in the field, that Cain rose up against Abel his brother, and slew him.

And the Lord said unto Cain, Where is Abel thy brother? And he said, I know not: Am I my brother's keeper?

And he said, What hast thou done? the voice of thy brother's blood crieth unto me from the ground.

And now art thou cursed from the earth, which hath opened her mouth to receive thy brother's blood from thy hand;

When thou tillest the ground, it shall not henceforth yield unto thee her strength; a fugitive and a vagabond shalt thou be in the earth...

So the corn which Cain planted yielded not, and the cities which Cain builded fell to ruin, and Cain wandered the earth but hid his face, fearing his brother may be avenged by his life. He was cast out from the tribe of Adam as they followed the earth and tended the flock. They multiplied to fill the earth as the Lord God commanded.

And Cain went out from the presence of the Lord, and dwelt in the land of Nod, on the east of Eden. And Cain knew his wife; and she conceived, and bare Enoch: and he builded a city, and called the name of the city, after the name of his son, Enoch.

Goyme

Goyme (Gho-ee-may), the eldest son of the tribal chief, Uctula (Ooghct-shula). He is the captain of the guard that provides defense for the enclave. Goyme waits sitting at the sheltered lower area of the fortress along the wooden post wall constructed to encircle the enclave. He hears yelling from one of the spotters watching from the suspended upper walkway that looks out beyond a clearing between the wooden fortress enclave and the forest beyond. Goyme races up the ladder to the catwalk and over to the man, who points out to Goyme what he observed.

"There in the thicket, along the creek. I saw 3 sea-men walking there. They dropped down into the creek-bed." Goyme whistles loudly and yells to his lieutenant, "Take 12 men and go down through the back gate to the creek bed and pick them up. Don't let them get through; these are the invaders from the sea."

The lieutenant complies and calls his men. They race through a back gate and immediately drop down into a trench that is cut through to the stream right in front of where the bandits were heading. Half of the men setup on a bluff above the creek-bed, while the others lay low in the bottom, hiding and waiting for the men to show up. When they have made their way toward Goyme's waiting men, they are completely overtaken and although outnumbered, they still put up a fight and 2 of them are wounded. One by a war-club to the back, and the other with an arrow in the throat. There were five of them altogether, and they were carrying the closed woven baskets full of bounty that bore the markings of Goyme's tribe.

By the time Goyme arrived on the scene, it was pretty much over, and he inspected the captives who by that time had been bound with leather cords. The invaders were all restrained, as was their leader, who was held in a prone position with his hands and feat bound, and with leather cords wrapped around his neck, and tied to his waist in front, in such a fashion that he could not stand up straight but was prone, and not able to walk very well. Goyme took them back to the enclave, and presented them to his father Uctula the chief. Uctula gestured to the smiths who were busy at their craft stations, and they came and began attaching cords to the mens waists and feet, which they used to hoist them upside down, hanging like meat drying out in the open.

When Togtu's group gets back to the tribal fortress, the wounded man Bartu, who is limping and being helped by one of the others, sees that these bandits have already been captured and are hanging upside down and awaiting their fate. Bartu then looks the man that shot him in the eye again, but this time with a smile on his face, while the bandit has the look of failure and frustration mixed with fear, as he looks on to 2 of his captors that are preparing knives for a serious purpose.

The tribe is quite jubilant with dancing and chanting, and music. Togtu approaches the tribal chief Uctula, and presents the remainder of the bounty, and notices that the 6 bags that were taken by the bandits have already come into the possession of the chief, obviously recovered by the chiefs efforts when his next in line, his son Goyme, encountered them, wounded several and captured most of them. Togtu wants favor and recognition from the chief as he attempts to make his way up the tribal ladder, so the loss and apparent recovery of some of his effort is exceedingly disappointing. He explains what has happened to the chief, hoping for credit for the stolen bags, but the chief is not persuaded by the assertions of Togtu. Ula notices what the situation is as well and objects because she led the other women as they foraged for 2 and a half days for the bounty, only to have the bandits make off with it.

She is quite upset and complains calling him unfair and accusing him of treating his loyal supporters as slaves, "Uctula, my father would have recognized what happened to us; how we risked our lives to gather the material, then he would have given us credit for the bags. As far back as I have known you, you have been unfair. You treat us as if we are not your family, but as those from outside. I wish he were still here, he'd make you do it."

Togtu hushes his wife who at this point has gotten somewhat mouthy, at which point the chief is moved to considerable anger and tells Togtu to mind her mouth pointing to the bandits that are hanging in the background.

He speaks not to his niece, but to Togtu, "Mind this woman's mouth, or would you and her like to join them. If you gathered the material, why did Goyme take them from invaders. Why is it that someone else had them, and my son took the bags from them, our enemies, and has presented them to me? If you want credit, you must bring the material and present it yourself."

In this way, Uctula was telling Togtu that failure is for the weak, so don't expect to be given credit for failure, and don't ask to be excused for a mistake.

Ula loudly speaks to Togtu in earshot of Uctula, "We should just take the bounty that remains and along with the others from your tribe, we should leave and form another clan. Let's see his son Goyme bring the big animals that you do."

Sensing Uctula is getting irritated, Togtu then instructs Ula, "Ula, sweet woman, you are everything to me, but you need to learn when to shut-it."

Togtu continues the negotiation and does not press the issue about the 6 stolen bags with the chief any more, but accepts the credit that Uctula has allowed. He and Ula then start to walk back to their shelter, and as she passes by the 5 men hanging there, Ula grabs a knife sticking out of a log, and uses it to slash the neck of the man that shot Bartu, causing his immediate exsanguination.

Lord Glanister

Civilization progress, and witness advance in complexity and institution, but has always most recently contemplated even the most primitive notions of divine right and governance burning beneath the feet of kings; always beckoning, whilst warning the Strong-Man in the delusion of his mind as he struggles against the lustful contest for his seat and head; always present about his door; fear of the usurper, the angry tribe, until providence gives solace, consoling..., redeeming..., democracy! Manifestly born amongst noble, but fostered amongst simple deliberation of tribe and not elite; whilst in the mind of noble, unsure of his hold on privilege, but which history and providence has bestowed his right. Assuredly now, superior and more purely virtuous order; the upper hand persists, albeit more demanding of gentle and righteous administration than the fore.

Assuredly, now tribe has small voice with comes contentment, allowing, demanding, gentle and righteous rule of king, ministers, council, and the elected amongst them in the new order, preserving and refining requisite privilege amongst noble. Surely, fostering and husbandry bestowed by superior order ensure virtue; to impart proper and right deliberation, which insolent, tribe would not produce. Invariably, a new order; the establishment and hybrid synthesis of common with elite functions not efficiently but for superior guidance from the initiated hand, to learn and perfect the authentic and secret craft of civilization, in theory and practice. Firmly now, noble is ever sought, exalted secure, tenable with hold atop; operating unbroken; now superior refined order, contrasting the brutish order.

New order, ministered through learned and skill, by practitioners, not beloved nor virtuous champions of old. There is naught to demonstrate in character, strength, nor valor, while tribe is beneficiary. The events of archaic time, the logic of circumstance, sufficient to ensure noble right; whilst, preserved and occasionally demonstrate the judicial use of power.

Small voice powers new order, giving contemptuous delusion to the tribe believing to oblige noble, they hold sway. Noble right, just as past, mocks the notion; instead, with new order and broad voice, having acquired tools to accelerate and perfect their delusion, perfects conspiracy. Having swelled the brethren, increased secure from the fore; whilst championed and true noble of old having righteously come upon his post, exist no more; only skilled ministers and reasoned practitioners of privilege and oppression. Gone, the popular true and sanctioned champion and sage, their memory a token symbol giving plausible cover for the evils of illegitimate keepers; merely members of a bygone cult which may only bear the title, 'elite-ist'.

§

England 1827, His Grace, Lord Phillip Bonefey Reginald Glanister, Duke of Coligston, his daughter Pippin, his wife, the Duchess, Blythe, his personal solicitor, H. (Hector) Peter Wellingham, of the board of the corporation, of which His Grace, Lord Glanister, is Chairman and the largest shareholder. The corporation is involved in the cotton trade with American growers, and has considerable Irish land holdings as security capital. He is in the parlor with the solicitor H.P. Wellingham, when he is informed of a visitor from Ireland, who is his agent-manager, left to oversee his lands; to manage them and collect the rents. He and his Solicitor retire to His Grace's study, and the man is sent in. He has arrived in a coach with wife and children who wait in the outer court, taking shelter in the stable, from the cold and rain, not feeling worthy, nor invited to enter a grand house.

The Irishman, Callaghan, clean but shabbily dressed in tattered clothing, but which are his best, enters. He tells the Lord, "Your Grace, having prudently subdivided the land many times, but there is not advantage for the poor decrepit souls to work or improve the land, and now the crops have failed. So you see, your grace, they are simply dying, not having meat nor corn, nor morsel to eat, and those which remain have moved headlong onto the highways with their meager belongings to find work in the houses, or in other lands. Now, having no need of eviction for dereliction of rent. The rents have diminished to a mere fragment of years fore, or failed altogether."

Angered, His Grace then rattles on a litany of insults about the dirty little people, "So you have come to beg my help feeding your tribe with whom the world would be made fortunate by their complete demise. The lot of you are trouble, and not to bother with formal consideration."

This moves the agent to reveal that he, his wife, and his children are among those starving, or have died in the famine, "Neither my dear Brigid nor myself having had as much as half a cake in the past 3 days, but have given all we had to keep the young ones from succumbing. I thought it prudent to have spent our remaining few pence to make the journey and inform His Grace, a virtuous soul indeed, of the situation and beg his assistance."

To which His Grace replies, "You have come to beg my assistance to my own affairs? I beg your pardon, but whence have I need of a man of your position to inform me to attend my own business? I suspect you are here for other reasons that are less forthcoming. State the business for which you have come."

Callaghan replies, "Indeed I have stated the total of my concern, Your Grace, and have nothing more to ask, nor offer."

The lord is not satisfied with the answer, so he takes offense and calls to have the man thrown out, and his parcel stripped, "Solicitor Wellingham, would I not be well within my right to have this man imprisoned, and his parcel stripped for conspiracy against my estate?

The Solicitor replies, "Indeed Your Grace, well within the letter and the spirit of the act."

His Grace continues, "There you have it.   Against my better judgement, I am feeling rather charitable. So, you may redeem your position upon return to my lands and a doubling of my due from the lot of your sordid country and kin. You may take your leave at once!"

The imposing irishman then walks to the entry and dispatches the guard with a dagger pulled from his boot, laying open his neck, and alarming the other 2 men. He then latches the large wooden entry doors with timbers from the lord's collection, tying them with cords from the curtains, then stands blocking the door.

The suggestion of questionable motive moves Callaghan to anger and to retort, "I had suspected a callous brute like Your Grace might so act; indifferent to what befalls those who toil and die in his service. I thought he might act in a manner unbecoming his position, as indeed he has, instead of taking compassion on those whose lands were stolen years before, and even now are occupied by brigands and tyrants in far off lands, and who now expend their substance to faithfully attend his grace's business, and fortune. And, you neither deserve, nor shall you receive such grace from the likes of me. You have murdered your last Irish."

He pulls the lord's own prize sword from the display and waves it about, testing its blade.   The solicitor freezes in fear of what may transpire, and stays fixed on the sofa, as one resigned to his fate. His grace, Lord Glanister retires to the large desk, and begins to mock the man, "I have dealt with your lot from my youth.   The gallows is nourished daily by your sort. There is always accommodation for a treasonous paddy. How daring you come in here demanding, and upsetting my household. You, and your race, are a threat to the welfare of an entire civilization, and to my house."

The Solicitor pleads with the Lord, "I beg your grace, do not antagonize the man, can you not see he is without sense?"

Lord Glanister continues to taunt, and retorts, "Nonsense, I will not hold my peace. This man has not the stomach, nor the stature to follow what he threatens; he merely postures. When it comes to it, his blood betrays him, and his heart gives way. His courage only comes from drink and he's had none; I suspect that fact may be more than a small part of this outburst. He only shrinks in the presence of nobility and hath not the substance for vengeance, only beggaring!"

The Solicitor points to the dying guard across the room, and says, "Beg pardon Your Grace, that does not seem mere posturing to me."

The Irishman replies to his grace, "And you Your Grace, you may redeem your child upon the life of my youngest, and your wife, the duchess, upon the life of the next. However, it remains to be seen if his grace may redeem his own life this day."

His Grace replies, "How daring you threaten my house, you are just as I have said, and have not the courage. You will swing by nightfall, you have my word on that!"

Callaghan walks about the room in an agitated but controlled manner, and continues, "It should come to his Grace's attention that the 2 youngest of my house have already succumbed, and the 3 remaining are week and frail, and quite in need of your grace's assistance, to which Your Grace is sorely obliged. I intend to collect my due now."

He stops in front of the desk where His Grace is seated, and drives the dagger deep and firmly into it through the center of the lord's right forearm, pinning it, then walks behind encircling him. The solicitor then gets up and attempts to reach the door, but can not get through the barrier of hardware latching the door. Callaghan, casually walks to where he falls prostrate, and drags him back to the sofa.

He throws him back onto it and warns, "Sir, my quarrel is not with you. I pray you, do not force my hand!"

He returns to Lord Glanister. Then he demands, and receives that his grace give patent and title to his Irish lands, whom the english had stolen centuries before, back to the rightful owners. He demands and receives the necessary letters of passage from his grace, which concludes their business. The Irishman gathers more cords to tie the solicitor, planning to stow him in the wardrobe but the man again bolts and attempts to flee and sound the alarm. Again Callaghan leads him back, pushing him down, but the solicitor will not yield, so he pins him through to the sofa.

Then removing the sword from the solicitor, he turns to Lord Glanister, and wiping back the perspiration and angst of a desperate man, who contemplates his own fate by the act that must follow, he speaks softly to His Grace, "Hell hath reserve for those who feed their soul with coin and the blood of innocents. I pray you take no comfort in knowing that you will leave this earth soon, and within the few moments, your wife and child shall follow close behind. You have my word on that…, Your Grace!"

He then strikes Lord Glanister down, and goes into the chamber of the lord's wife, and of his daughter and does send the both of them, making good on the bargain with His Grace.

He leaves by way of the 2nd floor balcony, and returns through the side back into the parlor. He helps himself to some of the valuable wares about in the manor, including quality garments and accoutrements, admiring himself in the looking-glass as he tries on various attire while in the lord's wardrobe; just payment for a lifetime of pain and theft. All told, a moderate horde could make many men rich beyond their dreams, but he only takes enough to pay passage and sustain his remaining family.

As he takes his leave he encounters several of the irish servants and informs them about his grace, "I am afraid the lord is indefinitely indisposed and will not soon be coming to supper, until the matter is sorted out. It seems he has come across great opportune fortune. He and the solicitor must attend to the matter with which I have informed them. He has instructed me to tell all of you that he is beset with the urgent matter and cannot be disturbed, and asked that the lot of you entertain his guests, until he is able. His Grace has also informed me, that, because of his great newfound fortune, he is feeling great charity and persuaded me to inform the servants that by the ninth hour, they should feel free to acquire whatever objects and attire fancy them, and they are all to take the next several weeks to journey to their own lands for holiday."

He pulls aside the head servant giving him the envelope containing the patent deeds of Irish land holdings, and asks that he return to the lord's lands in Ireland and present the deeds to the county magistrate.

Some of the servants remark to each other, "…that does not sound like his grace, has he gone daft?"

Others believe their own ears, and are convinced of the truth of the matter. They start to roam about grabbing objects and piling them in their quarters. The next day, the 50 or so servants of the manor virtually strip it and leave for their lands.

As Callaghan leaves, there are guests arriving for the evening soirée. The well dressed Irishman greets some of them and is quite respectful in a subtle but mocking way.

He then boards the lords waiting chariot, instructing the coachman, "I am about the business of his grace, Lord Glanister, the good lord. His grace has instructed you, taxi me to the port my good man. Have your man load my bags, post haste."

He is driven to the port where he boards a ship with his family minus the 2 of his children who have already died in the famine, whom he and his dear wife regret to abandon while in their meager but lovingly fashioned graves, left behind in Ireland. She is sad and weeping and begs her husband that they return to Ireland once more to say goodbye, fearing she will never again see their graves, before they set sail for America, but he insists they must go now, and promises her a return, and to move them to the families new home, just as soon as it is possible. He pays passage by way of the booty taken from the Lord's manor. The manor is virtually deserted, so His Grace, Lord Glanister is only discovered 3 days after, having missed important business matters.

The World's Most

Circa 2050, United States Presidential candidate Thomas (Tom) Fullerty is campaigning with some of his advisers and enjoying a considerable 'lead' in the so called polls that by this time were really a publicly consumed contrivance, so most were neither fooled, nor blinded to that fact. Elections, for the most part, were always a forgone conclusion and the outcomes never really a surprise, so they never turned out with a candidate elect other than one of the 2 pre-approved choices. This was so much the case that most people considered the process to be somewhat of a joke and a farce. Nonetheless, maybe due to tradition or ritual, there were still considerable numbers of the faithful that enjoyed the pageantry and suspense or excitement of the contest, and so, participated; after all, it was thought that someone needed to occupy the office of *"The World's*

Most Powerful Man", even if the choice was pre-selected from one of the 2 candidates approved by elitists.

When the election is concluded, lo and behold, candidate elect Fullerty is the declared winner. The pride and jubilation of such an accomplishment, after such a long and burdensome campaign has led the good candidate to believe it to be completely legitimate, and based upon the strength of his character, and public confidence in his ability to deal with the real problems facing the nation, and to some extent, the world at large; confidence, which he believes he has instilled in the nation; having just been elected *"the world's most powerful man."*

Capital

[MD Narrative]: There has never been something so misunderstood and the truth about which, is subject to so great a distortion and misappropriation of truth, than that of capital, and the production and use of capital, called capitalism. For many reasons having mostly to do with greed, the lust for power, and the envy political trade, a tremendous impulse has arisen to misinform the masses of the populations of the earth, mostly by those that have gained control of the capital of the world, based on their own nefarious motives. Ironically, they have preached that both capital, and the production of capital are evil and at the root of all the evil in the world, which is the very thing induced by the distortion. That is not an indictment of capital, nor capitalism. Both are simply tools, capable of being used for good or evil-exploitation. Capitalism is one of a very few things that separate humans from animals, the others being the use of spoken language; profound self-awareness; the ability to create; moral choice; and the capacity to know God.

The truth of what capital and capitalism is, and what they are not, should have long ago been understood and honestly discussed without fear of challenge nor distortion by the architects of disorder. So what is capital, and capitalism? Capital is something useful or valuable, employed to enhance the life standard of one person, tribe, or a whole nation; to add to the margin of safety against the elements of decay. Capitalism is a process of leveraging the capital resources which one does have, as limited as it may be, in order to create more of the same or another form of capital, thus enhancing one's own elevation in life, and by extension, most everyone else's. The creation of more by the process of leveraging is called gain or profit.

What capitalism is not, is the opposite of communism as the myth has been propagated. Capitalism is not a political system, as has been widely disseminated in the halls of academia and media channels for generations. They have propagated the conflation of the capital economic system with that of a political system. The 2 are separate and distinct, and really serve different purposes. Capitalism transcends time and politics in that political systems will have come and gone while capitalism remains, and is universal. Capitalism enables political systems. Like gravity, or ocean waves, it occurs naturally, and can not be altered to any appreciable degree.

Communists of old, of necessity, practiced capitalism and could not have survived without it, however, the opposite does not hold. As soon as communists were denied capital by those that practice it, or ceased to practice capitalism themselves, the communist political system failed, then ceased altogether. The cultivation and production of capital employed to enhance life and civilization is what is called capitalism and has been established since the foundation of civilization, many millennia in the past.

Ever since primitive man discovered that he could plant a few seeds of wild grass meticulously harvested by foraging, and carefully cultivate a crop, yielding many multiples of the same seed which are better formed and superior than those he started with, he has practiced the production of capital and profited. When he took one stone and struck it against another, breaking off pieces to shape it into a sharp tool, then used that tool to strip an animal of hide to clothe himself, or slice the hide into leather strips used for tying coverings over his feet, he has practiced capitalism. When he discovered that he could heat rocks and get metal, then mold the metal into a blade, then use the blade to hew trees, and create lumber, then use the lumber to build a house to shelter his family from the decay of the elements, he has practiced capitalism. When he refined and massively duplicated all of these innovations on a large scale, he greatly accelerated the practice of capitalism, and improved the civilization he was a part of.

The advent and development of civilization, which is the natural progression of humanity, could not have proceeded without the production of capital, it was simply not possible. Advanced expressions and extensions of the same process is what has created the world that we now live in; it absolutely would not have come about otherwise.

It may be lost upon some that there is more than one way capitalism is practiced. Modern societies have been swayed to practice a pure form of large scale group capitalism at the extreme, meaning, the individual is forced to surrender his own right to practice individual capitalism except to a very limited small scale, in order to allow for the creation of large monopoly pools of capital to be concentrated, and controlled by a limited number of the members of a society that purportedly represent the interests of the whole; they are the narrow-interest. The reality is, the narrow interests that have garnered the power to monopolize the control of capital, only represent their own interests, and use the power through the political system to secure their monopoly. The individual was more or less forced to abdicate his right to profit and create capital, or to so limit his involvement that it failed to adequately serve his need, and by this, the practice of the production of capital was largely transferred to the narrow-interest. This of course, was a corruption and diminishment of advantage by the individual, and thereby, the greater portion of the society as a whole. Instead the advantage only accrued to, and disproportionately benefits the narrow-interest.

The extreme example of the wholesale transfer of capitalism is that of communism. The production of capital was transferred entirely to the narrow-interest, and not just a limitation on the individual, so the result was a total shift of all the power of production to their control. To shore-up and maintain their concentrated advantage, they employed the power of their advantage in codifying law, to secure the advantage through the political system in what was termed the 'Rule of Law.' This again is not an indictment of capital nor capitalism, but instead, it is a reckoning of the evil side of human behavior and a corruption of universal law.

Prevailing World Order

[MD Narrative]: The glue that held socio-economic constructs together was simply the economic benefit derived. By the same principle, it follows that from the earliest of human history, tribes, clans, hunting bands, fiefdoms, principalities, villages, towns, cities, kingdoms, nations, empires, borders, organizations, corporations, global integration and governance and so forth; all are elements of the economic

constructs created around the drive for survival, more-so than any other thing. Some are a very simple expression of that drive, and others much more complex and advanced. They have encompassed the reigning political system, but generally significantly transcend it, whether leaning toward emphasis on individuals and a free capital based economy with liberty and value on the unique contributions by the individual, or more oppressive systems which only favored the elitists and narrow interests. This world view has been called the order of the day, or the word 'paradigm' also describes it.

§

Candidate elect Fullerty is fastly still enjoying his new status as POTUS elect, although the celebration, adulation, and incredibility of it all has started to subside, he still cannot believe that he has acquired the moniker, 'King of the World.' He has moved considerable amounts of his families essentials from the ranch in Kansas to Washington DC, and is planning on spending the weekend moving some items into the staging area designated for such, and spending some of the last free time with his family before he must pick up the mantle, to take-up the challenges at hand. He spends his days prayerfully contemplating the role, and basking in the glow and veneration of his family; his wife, his eldest daughter, along with the gleeful admiration of his 3 younger children.

After the appointed waiting period of transition, President elect and the Fullerty family bid a fond farewell to the folks back home and travel to the capital in anticipation of the move-in, to occupy the throne room and courtly chambers of power in the white house. He and the first lady elect are giddy as they walk the palace's corridors of power, variously discovering rooms and offices, and introducing themselves to the staff and the army of orchestrators, facilitators, movers, planners, etc. In fantasy, sometimes with humility and contrition, indulging, playing the part of the king in a pauper's disguise, as if he were just some common assigned to a task, and not really recognized or known; while other times demonstrating to an eager audience, his penchant for command, quick wit, thinking on his feet, and the art and skill of gentle persuasion and delegation.

President elect Fullerty and lady, walk into an area where several workmen and CHARLIEs' are busy moving furniture around and getting ready to paint and install decor into the area, when he introduces himself to them, "Excuse me, umm sir, and sirs, I am

President elect Fullerty, uh yes, and this is the first lady elect, Jeanine, umm, yes thank you. Can you tell me what it is you're doing? I was wondering where you got the authority to decorate, or whatever you're doing. The first lady elect has not been consulted yet, so please, stop now until we can sort this all out. Thank you!"

As the 2 walk away, men and machine continue as if the verbal dress down never happened, so he mumbles to the lady, "I can see we need to address the staff ASAP. ", she agrees.

He turns and sees a shadow, a man that walks up to the couple and introduces himself, "Mister President elect, ma'am, I would like to welcome the 2 of you, and introduce myself, my name is Steven Langdon, I am your chief of staff!"

President elect Fullerty replies, "Excuse me, you are who? My chief of staff? I'm sorry but I have never met you, and certainly did not appoint you to be white house chief of staff. You are mistaken my friend, my chief of staff is not yet here."

The Fullertys, feeling a bit uneasy turn and start to walk away, when the man moves ahead of them and says, "Sir, if you are referring to John Silva, then he is here."

At that the Fullertys stop and he asks, "John is here, where is he?"

The man replies, "I know this may seem a bit awkward for you, but I am your chief of staff. John is waiting with the others in the oval office; waiting for you to come and meet with a delegation. I have been sent to fetch the 2 of you for a meeting now. Sir there is a delegation awaiting your arrival. They would like you to come at once."

He replies, "Did John schedule this? I told him that I was not going to be bothered with protocol until after the first of the week. I need to get to the bottom of this. What's your name again?"

The man replies, "Steven Langdon sir, please if you will follow me, I will take you to where they are waiting."

Fullerty asks, "I thought you said it was in the oval office, I know where it is."

Langdon replies, "Yes, but I'm afraid you are not aware of which one they are in."

Fullerty asks, "Which one? Excuse me, what are you saying, are there more than one?"

He replies, "Yes sir there are, if you will follow me."

Fullerty objects again, "I'm sorry, but I do not know you or where we are going, and I am not going anywhere with you unless I am escorted by security." He sits down at a station in the corridor, and calls one of the CHARLIE units over, and instructs it, "Go and get one of the security officers for me, right away."

The CHARLIE unit answers him back, "Sir, I do not recognize you as my master, and can not comply with your request!"

Then Steven Langdon says to the CHARLIE unit, "It's ok, do as he says!"

The CHARLIE then complies and a security officer arrives. Fullerty then introduces himself to the officer, "Hello officer, boy am I glad to see you. I hope you are aware that I am president elect, Fullerty. I would like you to escort me and this gentleman to the oval office, apparently the other one."

The officer replies, "Yes sir, mister president elect Fullerty, I know who you are, and I will be happy to escort you and the first lady elect."

As they walk, Fullerty is somewhat agitated and defiantly stresses his displeasure to both the security officer, and Steven Langdon, "Was this meeting scheduled by the previous administration, or what? We were briefed, they still have until monday to engage in any last minute protocol? I was not told of this. I told John that I was planning on spending time with my wife and children on the last weekend, with a little quiet and informal get acquainted session with some of the staff, nothing more than that. Then spend a leisurely weekend relaxing and adjusting. Who the hell is it that thinks they can set time with me, let alone take me from my family at this late hour? I can see that I am going to have to lay down some rules, and soon. I am really surprised at John. He was my campaign manager. This is quite out of character for him!"

They proceed down a series of elevators, and several flights of stairs and through numerous long hallways, arriving at the designated oval office, the Fullertys enter and are astounded to find that the room is an exact copy of the one he toured before, down to the stains on the carpet, which are in the exact same place, color and shape as those in the other one, and are there by design. For a moment, the President elect, is a little disoriented, and feeling a bit light headed. He is not sure that he is not in the original room that he spent some time in during transition, and while on recent pre-inaugural visits, and meetings with various members of the outgoing administration staff. He mentally files away a question of how many actual oval offices there are, although still not altogether sure that there was not some sleight of hand at work. If it is a trick, it is elaborate because there are 12 individuals in the room when he arrives, some of them are definitely high publicly known figures. Those present range from a few members of the clergy in full regalia, to 2 famous persons that as far as he was aware, were deceased, and widely eulogized in the cloud-cast media, often. The others seemed to be a mix from media, industry, academia, and politics. He recognized Senator H. Bingham Stumble, and Senator Schuman, both of whom were in the media often because he was of the same party as Senator Schuman, and had occasion to have him speak in his support at a few recent campaign events.

The desk in the office is occupied by the man the Fullertys know as their campaign manager and intended chief of staff, John Silva. Upon seeing the man, he puts on the mantle of the commander in charge, and begins talking authoritatively in attempt to demonstrate his presidently-ness.

He greets the man and starts to demand what is going on, "Oh John, thank God you're here. I am a little confused, but I am glad you're here so that we can get down to some business. I need to lay down some rules, but I am afraid that this may not be the best time. I am at a disadvantage, because I do not know many in the room, nor what the purpose of this meeting is, nor who was responsible for calling it together."

The man at the presidential office desk says, "I suggest you sit down and shut up, mister president elect, sir. You are not in charge here! Does it look like you are in charge of this situation? No? So, big man now, 'King of the world', eh Mister El Presidenté, so what? I suggest you and the misses park it on the sofa, shut it and listen, and I'm not asking!"

Fullerty is taken aback and replies, "John, who the hell do you think you are, talking to me like that? You work for me, or at least you did. I may have a change of heart in that regard after what just transpired."

The Autonomous Economy I - The Strong Man

He is starting to get a little heated and huffy, as he starts to walk toward the door to leave the room, when the imposing figure of the security officer who escorted him there, stands in his way in front of the door with his chest heaving forward and arms crossed, with a stern look on his face, and the definite message, 'you are going no where, you had better do as you are told and sit down.'

The Fullertys comply and walk over and sit huddled together on a large overstuffed white sofa, next to the 2 members of the clergy present in the room. Fullerty then asks, "Please, John, tell us whats going on, because this is not what I expected upon arriving. Are we being kept here? What the hell is going on?"

The man at the desk then speaks, "Tom, you and Jeanine know me as John Silva, but that is not my real name. I am the man you knew as your campaign manager, and you did appoint me as your designated chief of staff, but that is all in the past. I am really sorry to have to break this to you, but things are really not as they appear, nor what you have been led to believe. Remember the canvasing episode? Do you really think you could have survived that on the strength of your charisma, or popularity..., 'finally a man who's heart is that of the people...,' hah, really? Do you really believe that you got here by anything you did? You were financed throughout, told what to do and how, every step of the way..., where to go, what to say, and what babies to kiss, were you not? I would have thought that you had figured that out by now, but whatever, some of us are slow learners, and that's ok. All will be made clear to you as you get oriented. What you need to know right now, is that you are needed in Wyoming ASAP. Your kids will be well taken care of while you are gone, you can contact them in the air, we need to leave in the next half hour to get you there. Your chief of staff, mister Langdon, will make the arrangements, and escort you to get you on your way."

CHAPTER FIFTEEN

Conquest

King Togtu

Late in the evening, after the lost bags, and the heated confrontation with the chief, Togtu is with his wife in their house and she says to him, "What I said to you and Uctula, that we should move to our own clan, I meant it. Why don't we do that? I know you Togtu, you are strong, maybe the strongest in the tribe. You should be the chief."

He explains his thoughts to his wife, "It is not easy to pick up and move off to start a new colony. Who will help build a new fortress and shelters? Who is there to help defend against an attack, and where will we get all of the necessary tools and supplies that now come from the tribe?"

He continues, "It seems wrong that Uctula has not been fair with credit for the bags, but he is correct, I failed to present them myself. He should have taken your effort into account though. We were attacked, and the bags were stolen by those Goyme captured. Those materials were clearly the ones Ula gathered. Its outrageous! However, picking up and starting all over is very difficult, and seems extreme."

Togtu has a very restless night, so the next day, having thought about it more he decides that it may be a good thing after-all. This is certainly not the first time he has been unfairly treated by Uctula. He is young, and his wife has no children to worry about. He also knows a very good spot that could be easily defended. He visited it often before he joined Ula's tribe. He decides to go there and take her that day.

They arrive at the spot which is about 8 miles from their current enclave, high up in a rocky outcrop. The location is close to the same river that flows by Ula's home tribe's enclave, and he asks her, "We can start here, and build in the rocks. We can get water from the river below. The place higher up will allow us to defend from a high position if we need to. My men will come if I ask them to. What do you think?"

She agrees, and they set about to move to the new location. It takes several months, but they recruit some who followed Togtu there originally from his tribe, and others from Ula's tribe. Togtu is popular, and many want to follow him. Some believe in him more than either Uctula or Goyme, whom they believe to be tyrants. They believe Togtu will be a fair chief. He is young and smart, and has been successful in his many hunting trips, bringing back considerable game for the tribe.

As time passes, the new settlement has erected a considerable enclave complete with a timber wall, gate, and catwalk just as the other one. It is higher up, and more naturally defended than the one Uctula's brother, Ula's father, had founded. They have also established trade with others from the area, so they are building some wealth, and continuing to expand. Togtu is chief, and he and Ula have had several children, so Ula is busy rearing them and no longer goes on forages, but instead stays in the enclave making use of her hands to turn materials into useful implements like leather and rope, and woven baskets.

Law of Conquest

[Narrative] All law invariably proceeds forth from the natural order of the universe. Some law, in remnant, proceeds from men, but inevitably, the foundations of that which rules the daily affairs of men are long established, and proceed from the nature of the world we live in, and are shaped as much by what is not, as by what is. On the one hand, the nature of civilization proceeds forth from the baser instincts, weakness, and limitations of the human condition, and on the lesser hand, it comes from the higher attributes of justice, hope, and love. Law is established by observing how things are ordained; the law stands because we are powerless to modify or move it, so it can-not be any other way. Men throughout history have always failed to act with sufficient compassion and grace, so the laws that govern

how we live with one another, for the most part, are ordained less by what has passed, and considerably more, by that which remains.

Empires have formed, and expanded over vast territories, up to and encompassing continents, by the brutal practice of war, invasion, and conquest. The historical record is always written around the affairs of the king or royal house, and chronicling of conquest is no exception. It is often an account of one tribe or empire being subjugated or conquered by another, mainly because the egos of the conquerers who commission the record, require the legacy be preserved, however biased or distorted the accounts describing the events may have become. Throughout the historical record, it is clear that the strong-man often engages in a dangerous game of sorts with his intended conquest, sometimes manifesting in treacherous overtures designed to lull the often intimidated victim into a false sense of safety and security, with assurances of peace through tribute, intermarriage, or the taking of hostages. Often the peace arrangement, or treaty, which may have been spurred, then arrived at by as little as the threat of invasion, or outright conquest by the more powerful aggressor, is structured around the simple payment of tribute, but often much more beyond that. Tribute, a word from the Latin, that means to divide the booty between the participating tribes, which indicates the ancient origin of the practice from its height, **"to the victor go the spoils."** More often though, the conquest is complete, and peace is kept through military invasion, occupation, and brutal oppression or outright annihilation of the vanquished, with complete and total subjugation by the conquerers.

It may surprise many people to find out that modern societies and their civilized systems of law and administration are built upon the very foundations established by the Law of Conquest, which simply states that the vanquished are subject to the supremacy of the conquerer; and given that the conquerers are those in control, they define any and all terms of the law. The vanquished have lost their previous right, and are now subject to those that own them. They may, and generally do, become slaves after some fashion, in some less than equitable arrangement. It may be bondage where they are chained and forced to perform labor to enrich the conquerers, or it may be that they are left to go about their business, but are subject to heavy taxation and the confiscation of their property, with some sort of political or military occupation; they generally loose political autonomy; they may even be removed from the land they occupied and sent elsewhere; or they may simply be separated from the flesh wholesale.

All of the worlds great civilizations, and many of the lesser ones that have existed up to the modern day, were established to one extent or another based upon the principles defined in the law of conquest. It was generally the basis for which the ruling elite rose to power and established and dictated daily life in tribal societies, which through proper and wise administration, grew to more powerful and larger tribes, nations, or entire civilizations. Tribal chieftains became kings, who in turn passed their prerogative on to their heirs, who in turn did the same, establishing long dynasties of reign. They employed large contingents of ministers in preservation, and ensuring the entire enterprise runs smooth. Many were the kings enforcers like military and police, or tax collectors; and judicial ministers, and lawyers charged with writing the kings will into laws, and advising the king on matters of law, and generally administering the kings laws and justice in regard to both domestic and foreign affairs.

High King Goyme

More time passes and Togtu's tribe has grown in size. Uctula has died leaving Goyme chief of his fathers tribe. Now there are 3 other minor tribes with separate enclaves under Togtu's protection in the area in addition to Togtu's and Goyme's. Togtu's tribe dominates all in the area including Goyme's tribe, by taking more of the game, and having instituted some agriculture, but Togtu is very fair with them and treats them all well, however, he has no diplomatic or trade relationship with the old enclave which he and Ula left. Goyme has heard of Togtu's success, and is jealous. He is obsessed with Togtu, and decides he wants to expand his territory by challenging Togtu, so he decides to go see his old friend at his enclave.

He sends a messenger to Togtu's enclave and receives word back that he is welcome to come for a visit. Togtu believes Goyme is interested in establishing trade with his tribe. He comes with his entourage, and is greeted warmly by his old friend. Togtu has prepared a feast and celebration for his old friend, and welcomes them in. They celebrate throughout the night.

Togtu, Goyme and a few men are sitting in the royal tribal house, when Togtu is suddenly feeling ill. He excuses himself, telling his guests that he will need to retire, but they will talk more in the morning. Togtu dies in the night.

The next day, Togtu's family finds him dead, and calls Togtu's first minister in to investigate what transpired. He then levels a charge that Goyme and his men have poisoned him. Goyme and his men take offense at the charge, then leave. Soon afterwards, Goyme and his men initiate a state of warfare against Togtu's federation of tribes, having killed their king, and gained inside knowledge of Togtu's tribal enclave, they have the advantage. They recruit some of the same invading sea-men that they formerly fought against, and come with a large force and overrun the enclave and the federation.

Goyme offers the participating sea-men some of the spoils of Togtu's territory and wealth for their help in the war, while he offers the vanquished tribesmen in Togtu's protection a deal that they may remain just as they were with their property and positions if they pledge their allegiance and tribute to Goyme. The ones that refuse are considered enemies. This includes Togtu's sons and his wife who is Goyme's cousin. Most are imprisoned unless they relent and change their minds, while some are executed. The male children of Togtu and Ula that have not escaped he has imprisoned, then executed. He holds his cousin Ula imprisoned for many years, but lets her out to join others when she is old.

Goyme has become the High-King of the area, which grows considerably over time. He achieves considerable wealth as a result. He has learned the lessons and value of ruthlessness; he trusts no one, not even his children, nor his ministers. He regularly has some of those close to him executed on frivolous charges, just to instill fear in the others. He endures many attempts on his life, and his throne. He lives a long time, and dies in his sleep an old man. His son succeeds him as king.

Right of Conquest

The Right of Conquest follows naturally from the Law of Conquest. It is the inevitable and consequential result of conquering, and being conquered, and pertains to the political status of both the conquerer, and the vanquished.

The right, which carries intrinsically genetic heritable conveyance to preserve it on the heirs of the conquers, may or may not be defined or codified in whatever legacy preservation accords are developed. The conquerers pass the right of conquest on to their progeny, while the children of the vanquished generally inherit their parents wretched state.

In modernity, in democratic arrangements, such preservation status and their accords are not generally widely known but may be confused with or mistaken for by wide social contracts such as a constitution, which can really only limit the administrative, sheltering the tribe from abuse, and outlines the relationship between the political institution needed for running the state, and the mass of the tribe, including the vanquished along with the heirs of conquest who hold the higher right, alike. In democratically structured systems, quite often the esoteric status of the right of conquest enjoyed by the heirs, only falling upon certain members of society and not the whole, may be purposefully obscured, while attesting that the more common right bestowed through publicly disseminated arrangements, is erroneously believed supreme.

It is obvious and requiring little substantiation, that the right of conquest is ended upon exercise and a new assertion of the law of conquest by another.

Hegemony, WY

Present day circa 2055, Hegemony, Wyoming, USA; located in the Teton valley, but somewhere outside of the visible spectrum of most of America; a ghost in the sense of existence. From the beginning, maintaining a cloud-blackout protocol on the city for the most part, it is well known in the rumor space of speculation and whispers, but it's actual existence has never been verified, which at this juncture is the exception to what is widely believed about information, and of course enhances the purpose for which its mystery and secrecy solicits. Obscure it through rumor; enhance confusion, and myth reinforces myth. It is at worst a wraith, and at best, a mirage, in an era when most things of this magnitude can readily be seen and believed.

The city was created using much of the technology of the day including massive numbers of CHARLIEs' requisitioned through various government protocols, husbanded by members of congress in the nations capital, on perpetual 'loan' to the city. It is closed to most everyone, so visiting the city is by invitation only. It is one of a series of like cities all around the globe, newly built by the core of what many have referred to as the Super-Minioned, Super-Class; the elite of the global elitists.

In the recent past, the Hegemonists would meet with groups of their ranks in private and public accommodations throughout the world, and always in very exotic locales, always characterized by exclusivity, which generally meant they were isolated from the crowds. The common person scares these people, and crowds of the common terrify them, whom they consider peons and rabble, and despise with great contempt.   With the rare exception of entertainment or performance talent, there are no common people present, not a single pedestrian soul, only CHARLIEs', and at the particular time, there is a rather sizable crowd of nearly 8000 persons from the elitist ranks that have come by invitation for the rituals and the sessions.

They began building a series of this form of private city around the globe, all of them named Hegemony. In addition to Hegemony Wyoming, their speculative-existence is believed to include in the mountains of Switzerland in Europe, the Australian continent, Asia, South America, and Africa. They are the new capitals, and fly the flag of conquest over the continents. Flags of domination, planted firmly over the entirety of the planet; designed and built on a new vision and model for the world, based on the determined shape of the future; of their own imagining, and favor.

All Hegemony cities are precisely laid out in geometric patterns, designed to fit themes of the particular religious order. Pyramids, obelisks, stars and pentagrams feature prominently in the layout and the architecture.   The focal apex of Hegemony, Wyoming is a trapezoidal shaped temple building atop the natural outcrop peak called Zion. Access to the top is by way of 2 roads dubbed, 'The way of the Serpent', which were so named based on the fact they wind in serpentine fashion, counter to one another, up the mountain. Both are paved from base to top with special tiles laid out in a pattern which makes the design resemble reptile scales. At the top, each road opens out into a large area resembling the snake's head. It is clear what the symbology represents. The city is spread out over 6 and a half square miles, and most of it is composed of a singular complex of connected buildings, punctuated with regularly intervening grand works of art, idol statuary, and magnificently appointed serene terraced courtyards and gardens. There are many streets and avenues with esoteric names like Hermes, Osiris, Horus, Isis, Lilith, Ashteroth, Baphomet, and so on. Transport about is via a permanent set of autonomous vehicles which never leave the area except to transport people to terminal facilities, or to nearby recreational areas. They are not CATs, but a privately owned fleet.

Rule of Conquest

The Rule of Conquest follows naturally from the Right of Conquest. It is the inevitable and consequential result of conquering, and being conquered, and pertains to the political status of both the conquerer, and the vanquished. The conquerer will generally have an absolute right to decree and enforce shades of any and all laws, policies, and courses necessary or which they may desire, while the vanquished generally loose most all of their previous right, and may have their rights in the new order redefined vis-à-vis the political economy. Those establishing the rule of conquest likewise pass the rule of conquest on to their progeny, while the children of the vanquished generally are those subject to the rule.

The rule of conquest is the basis for all rights despite the oft propagated misconception about natural rights; the belief that all persons are born with a set of in-born or inalienable rights to freedom, equality, and happiness. Nature enforces it's own order, and these are never naturally enforced, therefore these are not natural rights. Although ideal, this is all merely wishful thinking. Real, tangible, recognizable rights have always come by the application of force. Outside of the use of force, there are no rights.

It is likewise obvious and requiring little substantiation, that the rule of conquest is ended upon exercise and a new assertion of the law of conquest by another.

Regal Occasion

A ritual celebration of sorts takes place on the opening night of the gathering in Hegemony, and is every bit the cultic blood rite, with a large troupe of dark robed and hooded figures from the lower elitist ranks, acting a tradition in which, choreographically, they remove a mundane mask, revealing another more ominous mask underneath of something hideous; a creature from the abyss, or hades perhaps; some resemble gargoyles, or strange creatures from medieval stonework. This represents the true aspiration.   They then remove the second mask showing their human faces, without hesitation.   The ritual represents their hope to soon be able to work their dark and nefarious global agenda without having to hide or carry it out in secret.

There is music from an ensemble of musicians performing a lively set while guests arrive, and partake in the festivities, while a few of the minioned, Merudoch and Alaric, are watching and discussing the entertainment. Alaric is especially moved by the music, and shows his approval by light dancing and finger snapping.

He comments to his friend asking, "The things can build cities in a few weeks, or so I'm told, why is it that these fantastic machines have not learned to entertain nearly as well as humans?"

Merudoch replies, "Time, give it time, they will soon enough, as the era approaches! You are familiar with the doctrine of attainment? Remember, all things, nothing is left out!"

Alaric replies, "I'm sure you're right, and it is a shame and a waste because these guys are good.   Are they one timers, will the high council bring 'em back? Or are they out with the garbage as usual?"

He replies, "Why worry, you don't have a problem with that do you? What are a few of them now, when most will be a footnote soon enough?"

Alaric asks, "So you are convinced we go that path soon?"

He looks his friend sternly in the eye, and replies with great and authoritative emphasis on certainty, "Absolutely, absolutely, there is no other possible path! I'm worried that we are not going there as fast as we should, for safe margin. It's on the agenda."

At the conclusion of the nights celebration the musical ensemble are not seen again, and later there is a media report where they have been killed in a freak accident. The celebration rituals are teeming with priests or shaman of sorts, which are often involved in sacrificial rites. There are mountains of food, and exotic animals including birds and rare species, with some of them on the menu; rare artwork, and elaborate costumes. Great quantities of precious stones, and especially gold is everywhere, adorning everything. The bacchanal burns throughout the night with great revelry.

The next day, a plenary panel is in session. The regular planning with the full panel is something that takes place every 3 months, with the exception of unplanned contingencies. Those present are the who's who of the world's elitists, including those from industry, banking, the political establishment, entertainment, church and temple, and so on; all that have achieved extraordinary success as a result of their affiliation with this cult of global cabal.

With the pageantry befitting a royal coronation, or the anointing of a king as in a past era, a panel of robed and hooded men sitting high up in front of a large room of elitists, there are several involved in discussions as if the proceedings are in a court of law; a style called dialectic discourse, which they use for the process of discovery and decision making. They are discussing the political and economic state of the globe, and their particular individual and group holdings, territories and the resources they oversee, and the competition. Some squabble and fight, as petty and vindictive as one caught cheating at a game of cards. They complain, accuse, and threaten, how they are comparing to their SOB competitor in another country, and how unfair it is that the other realm has been given an advantage. They discuss what is being produced in which realm, and by whom; who their agents there are, and the trillions of profit being generated. Nary a word or mention of the misery such competitive manipulations wreck on the poor and hapless masses that suffer under the buffeting.

A man, Sargon, sitting in the upper level of the panel asks questions of the advocate arguing his position before the council, "Why are the natives so restless, have they been shown an example, and why has that not worked? Who are our agitators, and are they countering the problem? Further, what steps have been taken to deal with the troublemakers? Have we had an incident?"

The advocate for the Asian realm, Kahn, is in front of the panel and answers, "The problem stems from the Lanzhou incident.

Intrenched or maybe I should say insurgent interests are the root of the problem. They cropped up and organized afterwards, and have been quite stubborn and have proven quite difficult to dislodge. What may have been a fairly manageable problem early on has really grown unmanageable because of the incident that was supposed to solve the smaller difficulty. We are afraid any further action will make the situation worse than it already is! We are worried that further action may result in a number of deaths!"

Sargon continues inquiring, "Millions die every day, often at our hands. Tell us about your operation there? Have you employed the necessary resources to set a series of events in motion? If that is needed, will you do what it takes? If it wipes out a large portion of the population, what is that to you? Profit flow has diminished, and that is not acceptable."

The counter advocate, Alexander, speaks, "Your eminence, they have not so much as had an arbitration, let alone effected a properly designed incident, and they certainly have not put together the necessary resources to support a cascading series. They are sympathetic with the agitators that stir the waves. This thing has inept written all over it. I would recommend moving the resource to our operations in eastern Europe."

Sargon then takes counsel for a few moments with others on the panel, and gives advice and a decree, "Give them 20 days to reconcile the problem and if by then it has not been reconciled, make an example, and have the operation killed, then install new ones. It's no more complicated than that. You may take this under advisement, you have been warned. Do not let the shadow of this situation darken our business any more. So say all?"

The panel all raise their hands in the affirmative. Sargon continues, "So say all! So said, so agreed, so sealed, so done! Whats next?"

Economic Conquest

The ideas surrounding the advent of civilization evolved and later took root among people to modify the established order of things, to give more of a voice to the common of the tribe; to establish the right of the tribe to preserve him from abuse of the king. The ideas of democracy, though established by the ancient greeks eons before, never took hold until eventually when conditions in Europe were right. The tenets of democracy were to give the collective tribe more of a voice in their own affairs, and the affairs of the realm, which of necessity created a situation where elite kings and their lords had to take a step down from absolute power, and metaphorically replicate themselves, assuming lower absolute realms of power. The practical status of monarch has been divided, as it were, into many members usually represented by some body of ministers that are more deliberative than the one-man king. This same advent also gives rise to the voice, and thereby, the power of the tribe over the affairs of state. Now the tribe has voice and power in demand of policy and the nation as a whole; an assumed balance checked still by the abiding power of rulers from amongst the ranks of elitists.

Democracy, a modern boon to the tribe, has also given rise to new forms of conquest, specifically economic conquest in which the power of money is employed to bring a nation or tribe to its knees without firing a shot. The power has come about primarily due to the myopic role of many in the tribe, matched against the cunning machinations of those advanced in the art of deceit and treachery. Monopoly paper currency and monopoly banking; debasement of money; political and monetary favoritism; propaganda and the control of the perception of reality; complex rules engineered to favor one group above another regarding the formation and repayment of capital and credit; usury; cronyism and corruption through policy and favoritism; the employment of economic trade sanctions or blockades, which are boycotts designed to bring about capitulation or compliance by a so called 'rogue' nation; outright corruption of those charged with keeping the coin of the realm pure; on and on. All, weapons in the arsenal of the conquerer.

All are forms of economic conquest and yield the same result of subjugation and control as that of outright military occupation, with the exception that the ability to genetically pass on the right of conquest as in the past, has shifted to political control through henchmen and minions, to preserve un-broken, the autocratic cabal. For the victims of abuse, they reap untenable debt, inflation, debasement of economic power, the occasional raiding or collapse of accumulated wealth, destruction of a nations resolve, economic enslavement, poverty, destitution, the loss of productive capacity, and so on.

We may call this abuse a right as defined, but in fact it is not a right; merely an unjust abuse where it is difficult to inherit the right to perpetually steal from the masses of the tribe; it always tends toward failure of the system because it violates the certain universal laws of equity, and because the victims eventually see past the crown without a face and catch-on and stop the abuse through reform or outright revolution. The economic conquerers loose control through countering by their victims or circumstances, therefore this right is limited and more difficult to preserve than the rights which bloodline inheritance bestows; in this case blood is thicker than money. Nonetheless, in modern times, economic means have become favored against advanced civilizations, while the old-styled military invasion and conquest have carried on and are still widely practiced against underdeveloped tribes, although having declined.

Pinnacle of Power or Puppet

Along with other staff, Senators Schuman and Stumble are present accompanying the couple in the aircraft as they make their way to the city. When President elect Fullerty and his wife arrive at the midweek session, the 2 senators show them around and introduce them to protocol. The 2 men send the misses off with some of the other women present, to enjoy some of the festivities, while they pull him aside to answer some of his questions, and intend to set him straight on what is to be expected of him. There is a pressing issue that he will need to address soon, and the 2 men along with others have their work cut out to bring him up to speed. They are outside in the early evening while it is still light out, in the heart of the city, and the night is warm and clear. They are sitting at a few benches arranged nearby where the music and rituals are being played out.

A CHARLIE approaches and asks the group, "May I get you gentlemen anything?"

Both Schuman and Stumble answer asking for Scotch, while Fullerty replies, "Nothing for me, thank you!"

Senator Schuman offers an observation, "Tom, it must come as somewhat of a shock to learn that your election did not make you the *'worlds most powerful man'*, as its popularly reported. Then to learn there is an esoteric protocol and set of rules you need to abide by, but, I assure you, you will find it easier to adapt than what you may be thinking at this point."

Fullerty replies, "I'm not sure what to think at this point. I'm having a bit of a problem taking this all in. Answer a question for me, why me? If what Silva, or whoever he is, if what he said is true, why was I the one chosen? I, umm, I don't understand. Why, why didn't you choose someone already initiated into, ugh, whatever this is?"

Senator Stumble answers, "Some answers are not going to come easily. I am not sure anyone can answer that question. I don't know and Cy, you don't know either, do you?"

Schuman answers, "Nope, and it doesn't really matter either. What you have do deal with is the fact that you are the man now."

He replies, "Well, I'm not sure I bargained for this. I'm not sure my heart is in it. What if I don't want to go along with whatever this is?"

Both men reply, "Thats not an option!"

Schuman continues, "You are in it with both feet now Tom. There is no out. Just do what is expected of you, and enjoy the rewards."

He replies, "But I'm no more than a figure-head, a marionette; this is what they mean when they say you are a marionette, isn't it? I have no real power do I, I'm just a puppet, aren't I?"

Stumble says, "Tom, you just worry about doing what you are called upon to do, and everything will be fine, really, I mean that. Before you know it, you and the family will be having the time of your lives. There are tremendous perks, and unbelievable rewards. However, and I have to warn you, if you think you can contradict the mandate of those you work for; if you think you can compete in a contest of wills, or take any initiatives not approved; to the same degree that there are perks, there are penalties. Keep that in mind!"

Fullerty replies, "That sounds like a threat, I don't work well with threats!"

Schuman says, "Take it however you want, just remember what we told you, and you'll be fine."

Stumble says, "You are not really a free agent. We, the two of us, are free to an extent, but even we have to abide by some strictures. Your office is of considerable interest, and so a special case. I'm sure you can understand that. Your position is too critical, you are the one the nation looks to when we want to get their attention. I know you are familiar with the concept known as continuation of policy?"

He replies, "Of course! I campaigned on reforming several laws that have continued for way too long, to fix them. They are wrong."

Stumble continues, "Well, thats not really up to you now. What I meant is, have you ever wondered why seemingly bad policies continue on between different changes in the seat? The continuation of policy is necessary because many things take a considerable time, generations, to enact, due to political considerations. Time is important, to ensure we not get setback; not generate excessive opposition. So its necessary that some initiatives carry over successive administrations. You'll learn your place in the food chain soon enough, but to really get up to speed, and understand whats expected, you should consider that your role is to be like a composite of all, and I do mean all, of your predecessors. None of them were allowed, nor are you allowed to deviate from the policy projects in place; and new ones will be your responsibility to push to completion. Do that, and it will go well."

Fullerty asks again, "I keep coming back to the question, why I was chosen. Was there something special about me, or my family, or the state I come from, or was I just, umm, easy? I have no special knowledge of how these things work. I've never participated in any organizations or clubs or anything of interest to you. Don't you have people that could have more easily fit the role, that are already, umm, prepared, and willing to go along, to get along? God, I'm not sure I wanted to know what you've shown me. This is really hard to…"

He is shaking and visibly upset with a tear that has come to his eye. He sits silently for a few seconds looking down.

The men reply, "Again, it is a mystery!"

Schuman says, "I'm not sure, but I think there is something new afoot. There are some new developments, and new initiatives on the horizon."

Fullerty, red-faced, and with his voice cracking, and his eyes watering up says, "I feel like you have just given me a death sentence."

Senator Schuman continues, "Get a grip man. Did you think the job was a parade, there are hard things in this world, and I know you, you are stronger than this. Look, I know you are upset, and I want to let you go and enjoy yourself, but lets leave with this last thought; play-ball, and you can last as long as you want; remember the 5 terms of Henninger? Get out of line, and contemplate what happened to old senator Owens, and congressman Bagley from Utah, and president Smith."

He replies, "Smith? He died in his sleep while in office, didn't he?"

Stumble answers, "Well that is what was reported in the news. What do you think happened? Go, enjoy. Oh, make sure you are well rested for tomorrow afternoon. I understand there is a special issue being discussed, and you're up. You are going to play a role in resolving it. You need to be rested and alert because you'll be in the spotlight soon. Time to show what you're made of. Ok my friend."

The men part with Fullerty and leave him pale and shaking nervously. He sits down on a bench with his head in his hands staring at the ground for about 15 minutes when his wife comes back with a beverage in her hand and finds him sitting, "Honey, whats wrong? Not having a good time? I'm actually having a pretty good time."

She hugs him and takes his hand pulling him to his feet. He says to her, "Oh, nothing. I guess, the job has started, and I'm, just a little, overwhelmed. I'll be alright."

They walk off to enjoy the night.

Old Order Passing

The world, and the western world in particular was founded on certain principles, along with powerful individuals and interests that have asserted destructive notions that are entirely antithetical to reality as understood by the western mindset, which are fond of espousing unworkable, doomed, and self-destructive doctrines such as engineering societies based on the idea of egalitarianism, meaning that all men are equal, and should be guaranteed equal status. While it is within the realm of morality and ethics to debate the points of this argument, and while this may be an idealized notion of how human societies should work, it is certainly not anywhere within the scope of workable reality. Humans have never been equal in any measure whatsoever. The question of equality, at best, is a matter of perspective. If ones perspective is from the 50 thousand foot view, then as ants, we may all look equally insignificant, but at the perspective level that men live and civilizations operate, the differences are vast.

However, the question of how we should live and treat one another is an altogether different one. Some are born with strengths and advantages that make them 'more equal', and able to outperform others in many areas of life. Physical traits like attractiveness, intelligence, race, talent, and so forth; and others are either born with, or have acquired advantages like the bonds of family, and the strength that comes from it; notoriety, inherited right, fame, wealth, fortune, status or position, admiration, political or financial or industrial connections, and so forth. There is a great tendency to deny, or cry foul upon the advantage given of inherent traits that over millennia have developed and become intertwined in the fabric or physical essence of the tribe itself. Such playing of the hand dealt, has given rise to abilities and aptitude inextricably locked into the very DNA, producing knowledge and ability that exceeds others of tribes not so endowed. These shades of difference are exactly why men are unequivocally unequal, and that which entitles the few to live at a much higher standard than the common average, and dictates that many, many more, live at considerably less than that average.

In the past millennia, humans generally agreed upon, defended, and abided by socio-economic rules dictated by the laws of survival, in that it was necessary to respect the concept of private property, in order to create the incentives by which the more capable members of the tribe could benefit by the ownership and development or general exploitation of resources, by furthering human knowledge, and the establishment of the institutions of power, which in theory and in practice, benefits all members of that tribe, but also creates divisions between men. Eventually, there were some that have and some that do not have. This gave the stronger ones that have considerable power over the lives of everyone beneath their particular elevation in society to the extent that some extended the concept of property ownership to include title to their fellow men and women. This of course, created tremendous loathing and contempt for these elites.

What gave them the right to establish, let alone maintain the structure of inequality?

After all, if we want to believe even minimally in the ideas and principle embodied in the notions of equality, which can never exceed anything more than the simplicity of observing that we all came into the world equally frail and naked, and were all destined to leave equally stiff and horizontal; then, given the compelling force of ordination, by what universal law was inequality established, or the private ownership of anything originally bestowed upon individuals or tribes? On the flip side, by the same criteria, what universal law was to decree that all things be equitably distributed between the strong and the weak?

What universal law or principles gave one man the right to own or control the life of another, and inversely, what gave society in general the right to plunder the strong to take for themselves? Is one principled action morally superior to the other? Does the universe favor one position over the other because some may benefit disproportionately, while others suffer disproportionately to the norm? Shall force be employed to even the uneven, and balance the scale? Even if some believe in the idea of moral superiority in the bent toward an egalitarian credo and society, we can not escape the use of force, societies compelling control over the Strong Man to the benefit of the weaker members, believing when they bring about equality, they are purging society of its evils and ensuring the efficiency of the future.

Both sides of this equation carry the same moral weight, thus they cancel the moral question altogether. Both are sides of the same coin, and both require force to maintain; the former necessary while not benign, but essential nonetheless; and the latter quite evil, and destructive to all in its application and its aftermath. The force required to compel each side is always present in abundance, while the induced politic which emerges being the synthesis of the two. Both sides constitute shades of the socio-economic ordained range of options.

So what condition established so powerful a contest? What is the origin? From where was it first derived?

Was it because God in Genesis gave man dominion over the earth and everything in it?

"*...And God said,... let them have dominion over... all the earth, and over every creeping thing that creepeth upon the earth.*"

Or was it simply by the act of declaration, or making a claim, or the law of conquest,

"*I claim this in the name of the king*",

...or was it decided by the outcome of war between tribes, or stronger and weaker individuals?

Actually, it was all of these things, as pedestrian or sordid as some may seem, and very little by lofty principles, ideals or concepts; not by the wise and fair deliberation of councils of sages, nor noble quest of the hero. Things from the beginning, were acquired and passed on by down and dirty contest between all parties; one acquiring and the next stealing from the first. The declared winners were generally those willing to stoop to the lowest levels, cheating and killing their rivals, and they were applauded and worshiped and adored by the masses, all the while, they were feared and hated by the same; and they established the law, and are remembered, while the losers suffer ignominious obscurity.

Even those true-believer minions, leftists, utopianists, that generally shunned private property ownership as somehow demeaning of the human condition, have the conviction of an atheist in a foxhole. If they could ever get the tribe to agree to completely ban the ownership of private property, to bring down the Strong Man, deep down they absolutely fear the result will be much worse than the relatively diminished indignity suffered by them under the status quo.

The rules regarding the inequality between men had long previous been decided and vetted, and it was the way the natural order had ordained, established, and sanctioned; and as ugly and seemingly unfair as it is, it is the optimum, the best, the most workable system that can be employed, and anything else would result in the failure of the race.

[MD Narrative] There began a general realization by most, something important that heretofore had not been given much consideration, but was now becoming evident. Now we approach an era in which the survival directives that dictate human civil order have changed and no longer apply, with the role of property ownership and the order it bequeaths diminishing, destined for oblivion, while for the first time giving approach to real equality between men, without requisite force against the natural order, which had been previously enforced through degenerate ideology and policy. It is no longer necessary for men to compare themselves by measure of economic performance; each deriving his worth to the whole based on his ability to produce. Indeed, outside of this measure, men have become virtually equal.

The Autonomous Economy I - The Strong Man

Also diminishing is the requirement to defend against that corrupt endeavor without moral repudiation as engendered in the past; it came in its time; it happened with little controversy; it was as natural as living. The old rules tied directly to the survival of the tribe no longer apply, and the performance of the Strong Man along with his disproportionate prerogative have also disappeared. People are no longer able to protect their ownership of land and material things, nor their status in the economic order, and the incentive to do so diminishes rapidly.

Conquest of Survival

[MD Narrative]: Now what do you suppose happens when the necessity that drives the ordained socio-economic construct is removed? What is the new social order; the reigning order of all civilization, that of necessity must now emerge? What new order ensues?

If the organizing force and principle around which these constructs are formed, is the innate drive for survival, and the ultimate expression of that survival mode diminishes its own requisite existence, then what is the effect produced by the logical extreme of that expression? How has the human condition changed when the directive for the prevailing economic order begins to disappear? How is it possible that the economic directives that dictate survival could simply disappear? Will there not always be the need for humans to live and cooperate in groups; some variation on social order, with co-operation in the division of labor, economies of scale, and toleration for the uneven distribution of wealth and resources, inorder to survive? Will men not be required to continuously put their backs to the plow?

[MD Narrative]: Does not God say in Genesis, '...cursed is the ground for thy sake; in sorrow shalt thou eat of it all the days of thy life... Thorns also and thistles shall it bring forth to thee; and thou shalt eat the herb of the field... In the sweat of thy brow shalt thou eat bread, till thou return unto the ground; for out of it wast thou taken: for dust thou art, and unto dust shalt thou return'...?

If the organizing force and principle around the socio-economic constructs for survival were to disappear, could that be anything less than the actual conquest of survival itself?

CHAPTER SIXTEEN

Affluence

Mutual Patsy

[MD Narrative] There was a Robo-Gate or Charlie-Gate scandal, as it was reported by the cloud-media in the nations capital which no person really saw coming, consequently nobody would really have predicted the impact it created, so it's really hard to say why, but of the entire history of political scandals, this one was very prominent amongst a confluence of circumstances and events flowing together, the result of which produced a torrent of trouble and is most likely among several occurring around the same time to have been the actual catalyst that triggered and set in motion the transformation that occurred. That this seemingly commonplace and mundane insult, most would think likely to be overlooked, could foster such energy and drive such change, although, some would argue that the same course would have ensued regardless, and the same result was an inevitability and would have happened no matter what the catalyst, or lack thereof; nonetheless, it did happen, and that is the history we have to deal with.

§

Circa 2047, political elitists in the capital, for the previous decade or so are caught enriching themselves, having kept an army of **ARCH 600** and **CHARLIE** units for their exclusive use as personal servants. The whole affair started to unfold right about the time that industrial productivity really began to skyrocket due to the rollout of the new machines to industry and commerce. Consequently, the nation reached record levels of unemployment due to the displacement of workers by these new productivity machines replacing them, and not just in traditional manufacturing and heavy industry, but also the heavily populated light industry and the service sectors, so they are quite agitated with many of them staging protests and threatening violent revolt.

People find out that they, the insignificant peon class have been bearing the expense for this excess and insult after about 12 years of hiding it by both parties, and demand that the machines are made available to them for their pleasure as well. The lies and excuses from the criminal elitist ran the gamut from: *"The machines utilized are part of a security exercise by the military and nation's security services"*, to, *"The whole project is classified under national security as test of machine suitability in the general population and society at large. We were going to tell you and let you in on it when ready, but now you've spoiled the surprise."*

It started when Grady Hunt, bean-counter extraordinaire, and dutiful minion, returned from the delegation tour at Dalbots in '28, and gave his report to senator Schuman, who later hatched a plot to back channel requisition a few of the early machines, running $12 million each. He gets Senator H. Bingham Stumble to help, a method of criminal graft pioneered in Washington in which each provides a patsy for the other, a kind of mutually assured "prosecution", combined with plausible deniability, usable one against the other. Together, these two miscreants access the usual secret slush fund budgets and acquire the funds by cross pilfering from each others committees, but then again, this was nothing unusual. On the contrary, it was business as usual for the self-serving, stuffed, basted and glazed turkeys in the nations capital.

Eventually with their successful dodge of accountability, others of the political and cultural elitist began to see that this perk could be theirs as well for the extorting. The phenomena started in the nations capital, then spread to other capitals and other high, heavily-monied hot-spots. So CHARLIEs' started to make regular appearance in the high places in New York, London, Paris, and the other great metropolitan cities of the world. Anywhere the minioned-class went, these machines were seen. It became somewhat of a media curiosity, and considerable attention was paid to them by the fascinated common onlookers. No one questioned the rich having expensive toys that the common could not afford, that is, until the story of how some came to acquire them gratis, courtesy of common sweat, was exposed, then it took on a different dimension.

After the excuses seem to run thin and with record protest and media attention on the issue beginning to mount, the fat-cats saw it was not abating or going away, and wanting to pull their roasting bacon out of the fire, they relent and in quick order levy a tax on industry, funding a small purchase of machines for the public's use and amusement, which were employed in major cities around the country. This only wetted the appetite amongst the public. Beginning to form in their imaginations, was the possibility of employing them for their own advantage. They became the object of desire amongst many of the common as a profit center advantage in business while easing their own laborious and dreary lives.

The political class has always had a penchant for finding a way of sharing the graft with their constituency of underclass, where a little goes a long way. The belief that it could better atone for their sins, and, not that it mattered much, but if it did matter even a little, it was somewhat better for their political future, and more profitable somehow than sucking-up with finance or industry big-shots, with whom, rather as those aspiring to the minioned-classes, they reserved the right to raid now and again for massive taxes, whenever their iron-lock on the first group of constituents was threatened.

So, those professionally skilled in graft and corruption were now somewhat off the hook for the interim. Having been given a reprieve, they were hopeful their paltry fix would take effect and it would all soon blow-over so they could go back to their usual business of pilfering and stealing.

A few CHARLIEs now started to walk the streets of America. At first, people did not know what to think of them. The initial ones were granted to local and state governments and municipalities, but their use was left up to the authorities given their charge. The familiar pattern then began to emerge all over again, although now it was on a more or less local level with local corruption so the outrage and the correction was local as well, and much closer to the individual. Many were employed on work crews for cities, doing light clerical or service work, cleaning up trash, or civil project construction and maintenance, but on the weekends they went home with local bureaucrats for their personal use. The main point is, the typical citizen got a chance to see them up close, and interact with them, but just barely.

Although prior to this era, these machines were primarily intended for industrial and business use, and so that is where they tended to concentrate, their demand for use by the common household and small enterprise started to surge with the general public, mostly amongst the gadgeratti, early adaptors, small to medium business and entrepreneurs, and techno-fools, but also with average people driven by simple curiosity, or those needing domestic help.   There was another significant segment of the population that would not come near one for any reason, the trust factor just not up to par. This group required a significant amount of convincing and it would be many years before they could feel comfortable with the machines.

Enrollment in engineering and robotics courses surged around the world, in every country from the greatest to the least. Entrepreneurial efforts led to numerous innovations of advanced products in the field of automation and robotics. The Dalbots X-Pack™ open API was used by these entrepreneurs to create a plethora of ARCH 600 and CHARLIE add-on products mostly in the realm of software, exploiting the X-Pack's™ facility and development tools, which significantly rounded out the universe of ability for these units to an apex of very high utility.

Most corporations started with a few machines, and added more as their production ramped.   They started with blank X-Pack™ templates, and through the process of training, and apprenticeship of the machines, they compiled extensive databases of proprietary knowledge and skill, which through the X-Pack™ portability feature, allowed the machines to refine, exchange, combine, accumulate, and ultimately compile all of that knowledge and skill collectively, into single or multiple, commercial or proprietary X-Packs™, containing all of the accumulated and refined knowledge of each machine's respective experiences and exposures. Some of the X-Packs™ were sold under license to others, but much was retained in-house. Now full complements of knowledge and skill contained in a single X-Pack™ master, could be simultaneously deployed into regiments of other machines thus producing armies of workers, each of which can perform most any and all jobs, for which any other single machine was originally trained. Productivity exploded, and costs dropped.

A virtual cottage industry was born where a person very skilled in some discipline could pass on his or her knowledge or skill to a few machines, refine it, then open up shop selling the expertise. Mostly in the disciplines of labor and low-skill, but also higher disciplines such as manufacturing, critical machining, fabrication, services, and maintenance. Even housewives with specialized handcrafts skills like sewing and needlework; culinary expertise; arts and craft works, etc. Eventually, virtually anything that a human can do, a machine can be trained to do, so the expertise wound up in a X-Pack™ and available for sale.

It was comparable to the advent of the automobile, when Henry Ford, an extremely pragmatic human being figured out that even though his Model-T masterpiece added to the gilding and stature of the rich, it could really live up to its potential, and was much better suited in the hands of the common person. Its probable but not certain whether Ford foresaw the automobile age coming or not; nor if he was even able to discern that the industry was much more likely to multi-facet, intensify, and proliferate, driven by the hunger and innovation of the masses, but one thing is certain; if the automobile had remained something exclusively for the rich, the world would have remained a much darker and harsher place for the next century than it did because of Ford's vision, and the course he set in motion. So it is with most technological advances since Ford.

Thousands of Dalbots hardware units were ordered, so much so, that there began to be a backlog of up to 15 months for delivery. When the time was right, and the Dalbots offering had experienced some level of maturity and seen enough utilitarian gains by the advent of thousands of 3rd party X-Pack™ enhancements, Dalbots, Boirix, and crew then decided to employ their own machines in full autonomous implementation. Henry Ford had innovated the assembly line, and now Boirix Dalgleish was destined to go into history having innovated Geometric Autonomous Production.

Dalbots caught up to demand in short order, but by that time the demand had experienced exponential growth. Dalbots was pumping out over 10,000 units a week in October, and 100,000 a day by november the following year.

Low-Lifes in High Places
[MD Narrative] Who are the powerful men, and organizations in high places, and what becomes of them when a new paradigm comes of age?

The Autonomous Economy I - The Strong Man

§

Circa 2019, Congress passes the Automation and Productivity Act, and immediately there are several political opportunists on the driving edge steering its progress, which is nothing more than the usual coarse of political activities in the modern era. How we have gone from the first truly devastating crisis that came so close to sinking the ship of state then, and now having arrived at this sad state is a matter of the linear progression of time. How we arrived with the lowest historical level of human cognizance about the reality of life and the world in which we live, is the result of decades of bad people telling big lies, often, and to enough people willing to believe them. Why we arrived back at a place worse than the previous one a half generation before, is due to an agenda furthered by the great lies, told by people who are in reality also simply minions of principalities and powers in higher places.

Forward to the present, circa 2050. Hubris and arrogance of the minioned-class is monumental, which most of the common have come to accept, believing it to be a necessary evil with nothing that can be done to change it. Elitism, is a virtual religion, and features prominently, but not as readily understood by the masses, however, they are not ignorant either. The political class always having had contempt for the common Joe citizen, now flaunt their deeds and attitudes with reckless abandon, sensing the coming age in which the world will transform significantly, and they believe the change will favor them. Their belief is that when that age dawns, they will no longer be shackled by any forbearance, but will be free to indulge any whim or lust without a thought or concern of the lesser. Corruption, continuous insult, and outrageous provocation of the masses by global elitists with their lapdog political-minions is the normal order of things, so much so, that common people have been forced to make peace with the situation the way a farmer does after his barn has burnt to the ground; nothing to do but accept the loss.

Elitists seduce political representation, which is then swayed to ignore the will of those they ostensibly represent and instead focus on doing the work of those that seem to be able to protect their minioned status from any account. A popular sentiment, based on a russian proverb is quoted often, '*We pretend to elect them, and they pretend to represent us!*' Even the youngest of bureaucrats are considerably beyond the temptation to simply ignore the interests of the people that sent them to do a job, so instead, they focus most of their energy on, and attempt to emulate or gain access to the minioned-class's clubs, associations, and institutions, currying favor to get a share of the perks, and their continuous aristocratic, decadent, indulgent lifestyles, paid in large part with public wealth, to which their more senior political kin seem to have perpetual access. This they do by building relationships with those that seem to run and dominate the process. The talented movers are taught the fundamentals of the system, the lessons of which are in reality, a kind of guide for bilking ill-gotten wealth and privilege out of the public trough. The system perpetuates the lifestyle of the unaccountable who have not earned the status; or who believe they have earned it; or are otherwise entitled to it; just as is the nature of a parasite.

The power of the minioned-class to enact whatsoever they desire, has become immense regardless of what the public thinks. Large swatches of goydom are willing and enthusiastic consumers of the doctrines and policies of global human enslavement. They are tools of the ruse without their knowledge. They are conditioned through various forms of communication; through institutions, news and opinion and the popular culture, in-cloud and out. The popular culture dominates, and is the gateway to shaping the mind-set. Those so under the pall, have become ardent defenders and proponents of the hegemonists mind set, readily accepting the lies as truth, inviolate. From the least to the greatest, every person with an ambition above the average, whether it is politics, or religion, media, or industry; they all clamor for favor to do the work of the wizards that dictate reality. This is the fount from which the true-believing dutiful minions spring, however, they are still less than half of the population.

All disciplines involving the arbiters of personal or civil discourse, values, and behavior have been co-opted into the machine. They stem from politics, commerce, banking, education, industry, church and state, the lowest of political climbers, to captains of industry. There are virtually no foundations, endowments, trusts, organizations, nor institutions that have not been co-opted and controlled from within. Governors, senators, congressmen, and mayors; the lowest of the googly-eyed activists and organizers, to the highest pillars of stability and trust; bankers, financiers, owners, and CEOs of industry;

editors, researchers, reporters, pundits, writers, cloud-casters, keepers of the records, teachers, administrators and educational counsels, publishers and the rest of those given the role of indoctrinating the mind; producers, actors, directors, writers, distributors, and those that depict the values and lessons of the betrayal of civilization through imagery and story, washing the brain with poison and corrupting the soul.   The ordained mind-set of the minioned-class is insidious, it pervades everywhere, and cannot be escaped altogether, short of absolute solitude in the remotest parts of the planet.   Most great examples of them can be seen in and around the citadels, real or metaphoric, of power and privilege.   The language they speak, reveals the instilled mind-set.

The Lumignosi Group, and Lumignari - The Lumignari are individual members of this global multilevel group of the minioned-class at the highest levels.   It is equally believed they have come up from the ranks of other ostensibly civic groups, clubs, and associations; or that they are only eligible, having been born into the bloodline; neither theory is proven.   One thing is absolute; there are no memberships amongst the common people.   These people are very secretive, hiding their affiliations and affairs with any negative or rumored activities.   They instituted and maintain the 'Black Cloud' protocol veil of secrecy about their association with the group, and if asked about it, they deny its existence.   The organization is eons old, having evolved in various forms, infiltrating and absorbing numerous groups and organizations throughout millennia to the present form of name and manifestation.   It is of the purest form, and religious in nature, practicing esoteric rites from ancient schools in ancient times up to the present.   They are considered to be those at the highest level, which orders the affairs of the planet.   All members are sworn to a blood oath in which they will forfeit their lives, or worse if they reveal even an iota of information about the true nature or agenda of the group.   The same blood oaths generally apply to the myriad of subordinate organizations.

The Builder's-Bag Group - Global cabal of wealthy elitists that believe the world and everything in it belongs to them, and is theirs to order in any manner they see fit, regardless of the fallout or the billions of lives wrecked in the process.   Many are cross-membered Lumignari, but because their organization is less formal in religious observance and differently focused, they are more involved in the day to day affairs of the globe. At the core, the membership is quintessentially elitist, and generally believe they are making everything better for the world, because they are smarter and thus more capable. In addition, because they have arrived at a position of wealth and influence, this means somehow they are ordained by a higher power to that post. The reality is, they are small minded individuals that never learned the axioms, *'you are not your brother's keeper'*, and, *'Do not stick your nose into someone else's business uninvited.'* Many are simply true-believing minions; others, indulgent members of the caviar, brie and wine crowd, having failed at completely blinding their conscience they make the lives of others a cause for themselves, to assuage the guilt they carry, having blackened their souls with money and sorcery.

Gregori Sorrows - One of the elitists of The Builder's-Bag Group. An extraordinarily wealthy man without a conscience. His reputation lives up to his namesake. As an arbitrageur, he is known for staking out very speculative high-stakes short positions in severely compromised fiat currencies from over-extended nations indebted to the global banking cartel in such a way that if successful his position triggers a panic, which can cause the economies of entire nations to falter and teeter on the brink of collapse or even go over the edge in the fashion of Dominoes.

Mr. Sorrows always supports through his 'philanthropy', those causes that create the greatest amount of dependency and engender within the target nation, massive hordes of 'zombie' agents, who make outlandish and incredible demands of the host government and supporting tax base through unconscionable, 'principled' notions of human rights and dignity. These same 'zombie' hordes are conditioned to act in a fashion similar to that of a pack of Jackals to get their way. This is done in an effort to compromise and weaken the resolve of the nations and cause them to submit and seek aid from the elitist global cabal. Mr. Sorrows has raised this form of extortion to an art-form. It is mind-boggling to many that one man can hold an entire nation hostage in this fashion, but with dark help from within and without, it is accomplished.

Counsel on Foreign Entanglements - Similar group as the Builder's-Bag group with many cross membered, except these are in the US, mostly from media, academia, and industry, and are very eager and googly-eyed groupies, and wannabes of the Builder's-Bag Group. Most of them are the recruits from the ranks of the minions, agents, and true-believers, and provide the eyes, ears, and babble-brains, in the process of doing the bidding of the other group. Most of their activity involves dreaming up deceptive but politically plausible arguments and 'principled' policy papers, which make their way into the legislative branch of government, then rubber-stamped into law, which then allow the elitists the latitude to make war and distribute plunder; then make continent wide wholesale moves of people and resources to suit their ends, thus they build up the economy of one continent, nation, or people while they simultaneously destroy another. This group greases the skids for codifying, thus justifying the crimes of their masters.

The Counsel on Foreign Affairs - The former name of the now more accepted and forthright **Counsel on Foreign Entanglements,** also sometimes referred to as the **'Council on Foreign Dependency.'**

The Royal Institute of International Hegemony in Britain - Sister organization to the **Counsel on Foreign Entanglements** in the US.

The Revenue Enforcement Agency - Dedicated to ensuring there is no concentration of capital accumulated by any person or organization outside of those sanctioned of the minioned-class cabal. They are skilled at the depraved depths of outrage and theft, using the color of law in their deception, intimidation, and the virtually unlimited power of government, to achieve compliance, extorting all excess capital from those not authorized to accumulate anything north of base survival.

Federal Bureau for Instigation - FBI - Ostensibly an intelligence and law enforcement agency tasked with keeping the nation safe from foreign and domestic criminals, who may threaten the domestic peace and tranquility while operating on domestic soil. It was widely known that this organization actually works as one of several enforcer and buffer police forces between paranoid hegemonists and the masses of rabble who might decide to end their tyranny. During an earlier era and up to the present, the agency was regularly involved in executing staged incidents designed for public consumption, useful in furthering the hegemonist agenda, and in directing the publics attention and sentiment away from suspicion of hegemonists malevolence in their midst. The pretenses of protecting the nation had been long lost for decades.

Demagogic, and Publican political parties - The biggest special interests in existence, massively controlling the rivers of money that flows through the conduits of power. During the nearly contiguous political seasons, while busy jockeying for the crown, they and their many backers, minions, and proponents, act together as integral cogs in the larger machine. They give continuous public speeches attempting to convince a desperate population against the truths they already know. They propagate through media channels that their policy positions and 'principled convictions' and intent are the polar opposite of their opponents and either party's namesake. In reality, they operate through a mirrored duality; 2 sides of an insidious talisman; the 2 out of one, which precisely mirror the other party's namesake. Their convictions can better be characterized as polar exacts. Very few people listen to any of it because of the duplicitous, dark, and circus nature of the members and events, and even fewer vote.

These parties decide the outcomes of their contests themselves based on the will of their patrons, regardless of the vote or the will of the populace. One may ask why they bother to conduct public campaign; endeavoring to articulate plans and positions? The simple answer is that in their minds, having gone through the motions, as hollow as it is, it somehow justifies the official positions they hold and their decisions regarding the burdens laid on the masses which ostensibly sanction them, as agents for the minioned.

The Counsel on Perpetual Warfare - Subsidiary group of the Counsel on Foreign Entanglements. Their mission is their namesake and they do it well.

The Working Group for the Propagation and Perpetuation of Financial Crisis - Adjunct to, and subsidiary of the Counsel on Foreign Entanglements. They are mostly a policy organization or 'Think Tank' tasked with the responsibility for finding ways to keep the worlds population under perpetual financial distress and instability.

The Usury Institute, and the **National Usury Counsel** -

These exalted bodies are responsible for methods aimed at the enslavement of the world through dependency; through loan-sharking, or more formerly, 'predatory lending practices', under the auspices of relieving pain and suffering.

The Federal Money Hoard, or **Fed-Hoard** for short, whose chief is referred to as **The Chairman of the Hoard**. Corrupt and bankrupt national repository of paper nothingness. It is also called the central bank. The Fed-Hoard, periodically creates financial bubbles by floating mountains of worthless monopoly, or uncontested sponge currency that actually acts as a siphon, sucking the wealth out of the nation that circulates it, while the political guard-dogs that set it in motion protect the entire corrupt operation to this day. Since it is monopolized by decree of the minioned-classes, it has no competition and so its regular, periodic debasement, only hurts those subjected to its unaccountability, instead of those that do the debasing.

The minioned-class arrange for those closest to the Fed-Hoard to get first use, giving them the advantage over others that get it second or fifth hand, after it's lagging debasement has been rekconed in the general economy and its worth proportionately discounted. While the peon-class, having no legal-tender alternative, pay in blood for it daily. It is handed out like toilet paper to friendly institutions, such as banks and other monopoly globalist causes under a corrupt rationalization that these organizations, having plundered and gambled away too much from the general economy, now pose a 'systemic risk' to the entire house of cards, and so must be propped up or we all suffer the collapse, but somehow cascading collapse happens with regularity regardless with each of these 'too big to fail' institutions, then their minions are right there to buy the charred remnant pieces for pennies on the dollar.

The Hypegnos Press, and Hypegnos CloudCast Media - Adept agents of mass hypnosis, these are the talking heads that daily implant ludicrous suggestion as alternate reality. They present us with the basketful of issues decreed by the powers and their 'self appointed arbiters of public debate', pointy headed elitists that daily, publicly, ostensibly, present and debate those issues, not for the purpose of discovering truth or justice, and not for the purpose of entertainment as they claim, but simply for spoon-feeding sanctioned orthodoxy to the public; actors plying their trade for a naive audience. They purport to be the total physical embodiment of all wisdom and knowledge, and the fount of all truth regarding civilization. They erect the straw man, ostensibly presenting every possible rational argument for the consequences of the issues they themselves decree and set before us. In reality, all sanctioned views represent a narrowly defined set of ordained orthodox possibilities, all of which are to the extreme of the inclinations of humans, so nary represent their holistic or natural views, which are more in congruence with the lessons of history, the natural order, and the laws of God. Their media views are instead designed to abandon the observer into the false belief that the arguments put forward are comprehensive and exhaustive, and to deliberate alternative views outside of those sanctioned is pointless or 'kooky.'

Consumers of the charade are swayed without their own knowledge or objection by their own reasoning that, *'the answers must be among those presented or they do not exist, after all, why would anyone want to deceive us?'* The targets reasoning, *'are we not all in the same boat, and do we not all sink together?'*, and, *'how could they deceive us, we are smarter than that; surely they would not take us for fools?'*

They ordain then drive correct the acceptable conditioned education, and by exercise, invoke the correct conditioned response. They invoke and propagate Heterodox-lobotomy. They engage, then answer questions no-one asks. They measure then correct the response, which is demonstrated whenever they switch on the public target's pre-recorded minds, by posing orthodox policy questions, then narrowing the allowed response into the acceptable range, which always gets the usual properly regurgitated pre-recorded pablum response.

Russell Tyrone "Swish" or "Flush" and sometimes "Squish" Rimschott - Media super-star and blabber extraordinaire who made his way up from salesman status, daily reliving his former glory as a high school basketball stud to become a nationwide 'syndicated and opinionated' (his trade slogan) cloud-cast media show host. Master of misdirection and manipulation, he skillfully speaks the language of defeat. He and his aspirant apprentices use rhetoric to command an army of the misguided. He is the Pied Piper of tens of millions of the frustrated masses that believe he represents an alternative and 'correct' view, in opposition to the hegemonist mind-set. The reality is he is a willing tool playing the vital role as the loyal opposition. He is very accomplished and highly paid in his deception. In misdirection, he exposes then disguises the crimes of the hegemonist brotherhood as simple disagreement among political rivals. He cries foul on designated decoys for their cheating and underhanded deeds, never acknowledging the real source of mischief, while declaring and neutralizing any genuine opposition as futility. He is the leader of a brigade of wannabee-clone, media stalwarts, emulating and paying homage to him even as they vie to be the first to render his substance; to sort the booty as the heir-apparent when the 'eminent one' succumbs to his demise.

Those vying for his title include the likes of **Shane Shenanigan**, an absolute giant of intellectual vacuity if it ever existed. Like his mentor, he misdirects from the source of mischief. With shrill rhetoric, he feigns outrage at partisan misbehavior. He is fond of brow beating to death, any that dare show independent thought or views outside of the orthodox line as he understands it.

Will Smear, the master of innuendo and insult. Without a shred of tact nor decency, his outrageous candor is monumental. He feeds the emotional need in his audience for ruthlessly disparaging others whose views are not attuned with the orthodox. He is free to defame at will. As a sanctioned minion, he is absolutely shielded from any liability for slander, or libel, or plain bad taste. Without even a feigned pretense of decency, he relentlessly, openly ridicules the targets of his disdain, ostensibly in the name of entertainment, in a public forum.

Allen Johns. Johns is quite adroit in misdirection. He is skilled at telling outrageous and incredible stories, painting a picture that makes the machine of the brotherhood seem so monumentally powerful, so vile, and so fearful that it can not be opposed. This has 1 of 2 effects in the washed minds of the masses: One, it neutralizes any genuine opposition, *"Pay no attention to the little man behind the curtain!"* Who can fight something so complete, so insidious, and so powerful?

Secondly, it makes the whole system out to be way too incredible to be a reality, so that anyone that purports the existence of such a system can be easily labeled a kook or mentally deficient. If one were to believe the incredible picture as painted by Johns, one should check themselves into a mental institution or so it seems. Sometimes it is easier emotionally to ignore the monster in the room than to acknowledge that such evil exists. In either case, it helps give them cover to carry out their nefarious mandate.

Fostering such ignorance in the masses by these tricksters throughout the ages has given this evil cover to grow and become the monster so described by Johns, however, it can easily be destroyed by exercise of simple choice, if people would only do so, but they have chosen to allow it to remain.

Rimschott goes by his childhood moniker and others he has acquired, but often he is greeted by his college pet name "Squish", invoking the question, "Hey Squish, why do they call you Squish?"

In addition to those mentioned, over the decades there have been countless, featureless, vanilla characters involved in all aspects of the compliant-zombie business, all pitching the orthodox mind-set to the weary masses in a deafening chorus day-in and day-out. It has paid significant dividends, with nearly half the population of 900 million souls in America swallowing whole, even defending the lies told by this machine.

Sham Affluence

[MD Narrative] Circa 2048, These were strange times. On one hand, the minioned among us had sewn up the political state and sealed it tight as a drum, meaning they had very little opposition to any initiative and could do whatever they wished for the most part. On the other hand, they would on occasion resort to the kind of tactics that helped them monopolize the process as if it were still a contest. The average person paid very little attention to what was going on in the political system simply because they were so cynical about the process. It simply did not matter what they thought or said, the policies regarding the nation were de-facto off-limits to commoners, despite the propaganda to the contrary, telling everyone to be sure and vote. It was a sham.

§

One of the tactics which helped the minioned gain monopoly political status was to popularize, and therefore feed the envy-trade. The envy-trade was where members of a poor sub-class were brainwashed through massive media campaigns and institutional conditioning into believing they had rights to a minimum standard of living and the same right to wealth as anyone, regardless that they had not earned it, simply by the fact that they occupied space within the confines of the nation. It was the entitlement industry. They never bothered to answer the question of where these mysterious rights to the wealth of others came from, or on what basis they were bestowed on anyone; that was irrelevant as far as they were concerned. They were born here, or simply they were here, and either of those facts entitled them and was enough. This was political theater at its finest because the minioned would use these people to make elaborate and outrageous justifications for their asinine assertions. In this way they divided the non-minioned, common-class between the poor and the middle, and used the envious brainwashed poor as a powerful weapon against the middle. Divide, then conquer.

Historically, the middle-class had built their estates through generations by more or less legitimate means, because they had worked hard and leveraged their capital resources to make sure they were self-sufficient, and their rights were foundational, which came by way of inheritance. The poor on the other-hand, mostly racial minorities and immigrants, were never self-sufficient and really never learned what it took to create wealth or capital and had no prior claim of rights, but were a class of people completely dependent on the charity, generosity, magnanimity, stupidity, and hand-outs from others, thus the burgeoning welfare state that had so threatened the nation with collapse a half generation before. At the time, the collapse was pushed back and thrown off with reforms in order to save the nation; but now similar threat was back and stronger than ever, but aimed only at the rapidly diminishing middle. As usual, it held little threat to the minioned elitists. So the poor remained poor, and the middle were made poorer as a result. Those familiar with the bottom rungs of society, felt better about this because, as they say, misery loves company.

By this time, enforcement of many laws and the system of justice in general was starting to break down and become of little effect and ignored because of double standards and selective enforcement, so laws were routinely disregarded by large segments of the population. Criminality was rampant and accelerating. The nation was headed down the path to anarchy. However, laws still meant something to the common because they were still enforced upon them. If you were on the enforcement action end, the law was all too real, but for the minioned it meant nothing. In their estimation, laws were to keep the common rabble in line, not for them; but if they were not subject to the law the whole system was failing. It must be applied across the spectrum to have any deterrent. If the government did not obey their own laws, and exempted a certain class from prosecution, they could hardly expect the common citizenry to abide either. The minioned routinely were exempt from virtually any crime, which were numerous. For all intents and purposes, they were above the laws they enforced upon others.

The minioned used manipulative tactics to gain political ground by using their considerable power to play the poor off of the middle in an attempt to make all poor. They played a game known as 'King Maker', where they would take some very atypical example of the underclass, and help them become relatively much richer. It started as a popular cloud-cast television show and became a reality in practice. The contestants were asked to describe what they wanted more than anything in the world, or what would make their lives better and the show's wealthy participants, who were usually members of the celebrity minioned-class, were supposed to make it happen for them ostensibly for entertainment, but with a heavy political component. It was an ego boost for the minioned who felt the need to express magnanimity now and again. For both the wealthy participant and the show, it mocked the principle of charity for the contestant because it did not come from a place of love, compassion or genuine concern for the person, but was motivated by a perverted sense of dealing with guilt, and with a strong desire to degrade others and cause harm. It was also a crass perversion of principle, because it was not directed at genuine helpless or needy, but usually at undeserving low-life contestants; people that were poor as a consequences of the poor choices they made, their poor behavior, and their particular cultural orientation. The game and the show really had very little in the way of redeeming societal value, it was simply a vulgar game.

Quite often the contestant would ask for something like vengeance on someone, and at first it was against the rules, so they were told to choose otherwise, but some of the wealthy 'King Makers' decided that the requests for vengeance were just as fair as anything else and demanded the show open the rules up to virtually any kind of perverted or crass personal lust whatsoever. These episodes made for some very colorful entertainment, but the show lost a considerable amount of its luster at this turn of events. However, it still remained popular by a smaller but more bawdy crowd. If the person wanted material wealth or an object such as a house, money, or exotic vacation, they were asked to explain what was stopping them and were usually given some trite lesson in ruthlessness with platitudes such as, *'The world is your oyster, go out and take what you want'*, or *'Its better to just take what you want now, even if you have to apologize later, than to ask permission now and be denied.'* Sometimes this worked, but mostly it had very little effect. The shows sponsors would often supply the contestant the desired request within reason, but often it was too expensive, so when they were brought back onto the show, they had little or nothing to show, then they were further instructed on the use of the 'killer instinct.'

One memorable occasion and episode set a pattern that was popularly repeated. A contestant asked for a large expensive house and was told that if he wanted a house that he should just go out and find one he liked and claim it; just move in. He found one unoccupied and empty and he claimed it, but soon found that he was the subject of a swift and violent eviction by the owner, and a personal claim for damages. The shows celebrity participant who originally gave the advice to the contestant, then stepped in and used his considerable power to back off the law-suit and reverse the eviction, then secure a title for the man.

The owner was forced to completely relinquish title and suffered a total loss. It was more or less a mob enforcing its will on an innocent man; such drama. The wealthy contestant had connections and influence with the sham courts and many shyster lawyers, so could easily overwhelm the rightful owner. It was a good precursor to the subsequent blood-sports. The rich using their wealth to overwhelm a victim not resourceful enough to defend against an unprovoked onslaught for the purpose of entertainment and self aggrandizement. The victims were always sneered at by the vicious audience and made to look foolish and greedy by the show, as if they were the perpetrators of the crime.

It was very crass, depicting itself as titan defender of the sham-entitled little guy against a villainous middle-class rich man. In the case of the initial episode, the low-life contestant subsequently lost the house very quickly to foreclosure by the state for failing to pay any taxes, so the original owner bought his property back, but the message was sent and clear; if you want something you cant have by legitimate means, just take it and don't worry about fallout until it happens, then employ the services of someone that can defend your action. The nation as a whole had arrived at the mentality of the 3rd world from half a generation before, long before this sad turn of events, so it was not particularly earth shattering behavior.

The practice escalated to the general population, and wholesale theft became common and sanctioned if the thieves were able to get away with it. It became a numbers game. In most cases, assets in the form of empty or unoccupied commercial or residential buildings, land, or houses, small business, large valuable equipment, even ARCHIEs and CHARLIEs; anything amenable to the shyster practice for transferable assets were targeted for this type of scheme. Individuals or groups would show up on parcels and erect a simple structure on them, or sometimes they would tear down an existing structure or remodel it, then go down to the county court and file titles, claims, homesteads and other bogus documents claiming ownership. It became theirs unless they were challenged by owners who were often unaware.

Some of the more creative groups of organized criminals employed sophisticated know-how and techniques to fraudulently access and alter public and official title records. They altered plat-map documents to subdivide lots, or changed legal owners, creating phony deeds and sale documents, then implant the falsified deeds back into the official records to cheat owners out of their property. Most would use unscrupulous lawyers to fraudulently evict legal owners. When the groups encountered legal resistance, they were assisted by various anarchist and leftist organizations who were well connected with the minioned-class, so powerful assistance was forthcoming.

It was all part of a bigger political project; the objective was to set a movement in motion, to accelerate the wholesale alteration of the considerably degraded political status quo from that point forward. The motivation and effect was to use the poor as a weapon against the middle in an attempt to dismantle the small remaining middle-class, and thus extinguish their remaining political influence, which had considerably diminished already before that time.

The spectacle was popular, so it played out in front of most of the poorer classes and further implanted in their minds that they were somehow victims of other peoples wealth, thus, wholesale thefts were justified in their minds to rectify the situation. Large segments of that population, for generations, had already been programmed with the mentality of the envy-trade, so were highly motivated to take advantage of the situation. They rarely considered who their master was, who set them upon such an evil course and what might befall them when all was said and done. They were simply a mindless devouring hoard, consuming everything in their path, while the middle-class were the actual victims and could do little other than stand back and watch the destruction of their world.

The practice of fraudulent record-theft gained considerable momentum before many jurisdictions cracked down on it, and property owners got wise using legal means to lock down their documents and their property rights. Once the genie was out of the bottle, it was extremely difficult to put it back in, so the practice hardly slowed but continued unabated. The crime became more sophisticated, as it was driven by greed, so attracted and employed high level help. It was inspired and fueled mostly by political motivation and greed, so little was done to abate it by criminal prosecution. Many jurisdictions cared little who owned the property as long as the taxes were paid, so the owner-victims with little resources were just out of luck. If your property was highjacked you had to deal with it yourself.

However, some fought back. Often the real owner would wait until the thieves were gone, and sometimes they would arm themselves and just barge right in. They would simply toss out the thieves and take back possession then correct the paperwork, or use the same resources as the criminals used to make it 'legal' again. However, they were still in danger of running afoul of powerful political advocacy action groups, but sometimes entire neighborhoods would themselves evict the unwelcome freeloaders and either destroy the property or present it back to the owners. Small wars with assaults and murders were not uncommon during this era. It was much like an older era in the uncivilized west when there was no law, except in this case, there was law, but it sided with the criminals.

In many cases, they went after affluent people that had the wherewithal to protect their property. This was a particular sport for them. They might choose the biggest house in the swankiest neighborhood to target, thumbing their nose at the owner. In many of these cases there was a powerful minioned entity of significantly greater influence behind the action of the interlopers, and it was often done with a plan of vengeance or a power play of sorts.

§

[MD Narrative] My parents were victims of the theft of land by on old employee, Rosa Cárdenas, who they had fired from service as a housekeeper nearly 2 decades before, for the criminal abuse of their children. She commissioned a house on their property and it was built by CHARLIEs while they were away for a few weeks. They came home to find half of their lot had been occupied by her house and not the whole thing, which was uncommon because this had happened many times before by the same type of people, only in most of those cases, the old house had been destroyed in the process to make room for the new house and the new occupant. However in our case, Boirix had instituted a directive of the highest order which should have protected all of us from any machine that may have malfunctioned, or that may have been instructed to do us harm in any way. The directive was total in scope, so finding Rosa the crazy nanny had built in his domain within two hundred feet was very unnerving, but at the same time a comical mystery considering who she was. Seemingly she did not really pose a threat, but nonetheless, here she was and it was a concern. The fact that she used CHARLIEs'; very expensive machines, and that they were now staying with her was particularly mysterious, so it was my opinion that she was connected and being used as a tool by someone that wanted to get close to Boirix. He did not share that view so he did not take adequate precautions like he should have for someone in his position, and that is the tragedy.

Scrap Metal

[MD Narrative] I have witnessed the transportation paradigm transform from one where people manually controlled the vehicles that transported them to and fro, to one where the transport itself became an autonomous machine with all the intelligence to not only move them where they wanted to go, at their whim, but do it flawlessly and at considerably improved safety, speed, and efficiency. The only problem is that the automobile was born from a culture in the United States and Europe that had gotten used to moving wherever their journey took them, as fast as they could get there. From the horse and wagon, to locomotives and steamships, to automobiles and airplanes, they always wanted to be in the front at the controls,

with a top down view, beholding all that lay before them, commanding tons of metal behind them, with power and control over their world; the sense of which was intoxicating. It became almost as if it was part of their DNA. They passed it on to their heirs who in turn were born craving the same power. Getting a license to operate a vehicle became a rite of passage, and they were hard pressed to ever relinquish the experience and embrace the future. Most did, but not all.

§

It's a small group with whom the culture of transport control and power was never relinquished or dissipated and they refused to give in, and to this day they cling to their cherished machines which have only gotten more sophisticated, with the technology advanced by decades over the obsolescence of the previous state of the art. Enhanced by machine labor which produces astonishing examples of homage to the automobile of old which now greatly magnifies the sensation of power and control.

Following closely in the stream of autonomous junk, the CATs, Jim Humphries feels trapped. He is driving his vintage 2026 Ferrari Testerossa, or '*Testosterrosa*' as Jim calls it, through the stream of CATs down the 46 lanes on the artery, and feels a very arrogant contempt toward the automaton yokels that surround him as they go their way in an endless stream of what he considers cookie cutter clutter. The advent of the autonomous vehicle is an anathema to Jim who is 40, and frustrated. It has no soul; there is no expression of one's self; no extension of his ego nor any part of his anatomy as was the case in the not too distant past when people outright owned their cars.

Back then they chose the particular make and model that seemed to fit their personality. It was an extension of their ego and an expression of their individuality, but now it has lost all meaning; no more than a pair of shoes. People don't own cars now per se because it is much easier to buy a monthly pass from the CAT fleet operators, and get transport when and wherever they wish; and there is talk that the fees will eventually disappear altogether, so the independent automobile dealers disappeared decades ago. However, there is an alternative and Jim has taken advantage, he belongs to a kind of brotherhood of enthusiasts that still hold to the credo of control; the feeling of the wind in your hair as you command the wheel and the accelerator. He is practical, so he uses the CATs whenever he needs routine transportation, but when he feels the need to indulge his passion, its his vintage Ferrari and his home-built beast. There are very few open parking facilities anymore. There hasn't been a parking lot anywhere nearby for 15 years. There was at one time, but since the fleets of CATs, they have all disappeared; mostly built over into more useful structures or parks and such.

On this particular day, he cant wait to get back home where he is very anxious to take his machine out on the track to try out the new mods installed by his CHARLIE Frank. Frank has special knowledge of Hydro-Ceramic-Metal Combustion Fusion technology (HCMC-Fusion), or mixing hydrogen, ceramic material, and engineered metal alloys, along with other materials into volatile gasses during the fuel metering and injection process which upon combustion, ignites a low level nuclear fusion reaction, yielding a very high level energy pulse, except without radiation. It was developed in the previous decades by fuel researchers looking to maximize the extractable power available from the old fashioned combustion of cheap ordinary fuels, when they realized the massive energy enhancement that a fusion reaction would yield when using combustion as a catalyst to ignite it. The technology gained only limited use and was threatened with oblivion due to commercialization of simpler and much cleaner, cold, and focused fusion reactors, along with breakthroughs in zero-point energy, several of which such technologies came to dominate the bulk of the energy industry.

HCMC-Fusion is dirty and requires a thorough cleaning of the engine internals very frequently, not to mention the requirement for extreme structural integrity of the engine components to withstand the energy levels. A few people were killed by engine explosions before it was considered safe, by developing and retrofitting special components to handle the power. HCMCF was relegated to very large machines which require up to 100,000 horsepower within a relatively small form-factor, such as large high speed earth drillers and such.

Frank's HCMCF expert kernel comes from a Kno-Pack™ installed, which was compiled from a lab where development work on the technology was done over five years, so he is very familiar with its characteristics and requirements, and Jim has directed Frank to install 2 power units on his home-built baby which he calls the beast. It is a vehicle of his own design, constructed with the help of Frank.

The Autonomous Economy I - The Strong Man

On this particular day, as he crawls down the artery in his Ferrari, he is moving much slower than he thinks people should move at only 93 miles an hour, when to his astonishment he hears the nearly forgotten sound of a horn blasting, bomp, bomp. He looks around but does not see where it is coming from, which is surprising because in a sea of sameness, it should be glaringly apparent. If it is what he suspects it should be very easy to spot, but he can't identify it; bomp, bomp, bomp; there it is again. Glancing up in his rear view, he finally sees it; another vintage old-fashioned automobile; he can hardly believe his eyes. Could this really be a fellow traveler? He slows and pulls hard to the left quickly cutting a swath through the swarm, across 16 lanes to the 3rd from the inside of the 23 lane artery. Sure enough; there it is pulling up along side of him. With windows cranking down on both cars, it's a dude!

"Hey bud, I like your car.", yells the dude, "I could not engage your com. Is it switched off?"

Yelling back, Jim brags, "Thanks, you should see what I have in my garage. This machine has a very outdated com, and no its not switched off, its just old. The car is all original factory."

"Are you in the group?", Jim shouts back, referring to the vintage automobile enthusiast club he is a member of.

The dude asks, "Well what is it, could I come over and see it?"

"Not today friend. I'm busy", replies a reluctant Jim Humphries.

He's a little annoyed that the dude ignored his query about the vintage auto club and seemed to brush it off, and besides he doesn't know this guy from Adam and he's not generally that anxious to entertain the strange and dubious characters one might encounter on the artery at 90 miles an hour. Jim starts to roll his window up and cuts hard inbound to the center.

As Jim pulls away, the dude yells, "I'm going to come by and see you, and you can show me what you got!"

Jim facetiously replies, "OK dude!"

Jim exits the artery at the next ramp and winds down the boulevards toward home. About a half hour later, Jim is outside his house and busy with his two CHARLIEs wiping his 'Testosterossa' down, hand buffing it tenderly, when the same dude he met on the artery drives up in his 2024 Rolls Silver Cloud. The Rolls is not nearly as tenderly cared for as Jim's Ferrari. Nonetheless it is still an impressive example of past automotive beauty and grace. The dude gets out and approaches Jim with his hand extended.

"Hi, names Steve!"

Jim extends his hand, "Glad to meet you Steve, I see you found me. I'm Jim."

"Yea it weren't hard, you were easy to find. My friend's CHARLIE found you in two shakes. I just asked him to look for Ferrari, and you came up. My nav did the rest."

Jim asks, "Well what can I do for you Steve?"

"Well Jim…, can I call you Jim? Seeing you out on the artery 20 back, uh…, I don't see too many like you now a days, so I thought I should find out what you're all about. You part of a club or something?"

Jim replies, "So you just followed me huh?"

"Hope you don't mind?", says Steve.

Jim, trying to get Steve to take a hint and leave, replies, "Well like I said, I'm kinda busy now, maybe you could come back?"

"What you got, in there", asks Steve, pointing and walking to what appears to be a still functioning garage like in the old days.

"Hey how about a beer? He'll get it and be right back", asks Jim, figuring, "The guy is not going to take the hint, so I might as well be hospitable."

Jim gestures with his thumb and little fingers extended to his CHARLIE with a sign like something tipping to his mouth, and 2 fingers raised on his other hand.

"Brewski, never turn one down.", replies Steve.

The CHARLIE returns moments later in butler fashion carrying a silver tray with 2 frosted mugs and 2 beer bottles, and a towel draped over his arm. The CHARLIE begins to pour the beer, "I'll take mine out of the bottle if you don't mind!", Steve interrupts as he forcefully reaches over and grabs the un-poured bottle from the hand of the CHARLIE.

Jim Jokes, "Hey be careful, that one is sensitive, you might hurt his feelings, ha ha."

Steve being dramatic and with a grimaced look of horror on his face says, "Don't much care for them damn things, don't trust 'em. I had a bad experience with one as a child. Ever since then, Ive been pretty shy of them. I never wanted one of my own."

"Well mine are ok, they never hurt me any. Besides they do come in handy", Jim replies.

Sensing Jim is not eager to unveil what he has in his garage, Steve attempts to goad him by doing a little bragging and challenging his ego, "I can beat anything in the quarter in that."

Steve points at his ride. The Rolls is positioned with the sun to the back of it, and the rays of heat are clearly visible rising off the hood. Jim senses that it had not been restored to factory condition, and begins to walk toward the car and toward the back side where upon he sees that Steve has modified it with large fat and wide slicks like a dragster or funny car of the past. The rear end also appears to be jacked up so that it is higher in the back. Jim smirks and giggles a little under his breath.

Steve says, "The motor is not stock, I have an LJ-RS24 under there. I had to widen the cage and remove the fender wells, then move the firewall back 8 inches to get her in there, but she's in there now. I built her myself."

He is referring to exotic Rotary Star engines. In this case, a 24 cylinder modular expandable engine with 4, 6-cylinders stacked-star rotary crank blocks like those of antique aircraft. They could be expanded to any size by bolting on another slice, or stage, rotated by 30 degree from the previous. Each block's shaft is piloted and keyed to accept succeeding stages. They normally burn a combination of Kerosene and Nitro, but were capable of running on plain old 93 octane. They were produced in limited quantity by Lyman and Jeffords back in the mid 30s. Back a few years, it was not unusual to see as many as 10 units trained together making a 60 cylinder engine that could develop in excess of 30,000 horsepower. They are still in use in large earth shapers and ships, and the like. Steve's rig was unusual to say the least, but Jim began to wonder if raw power was really the key to moving fast? He sensed a stupid challenge was forthcoming, and was up for it.

Steve brags, "She even beats them fixed electrics off the line, mainly because I get traction, but I can out accelerate them, usually. I had her out in the desert for a rally just last week after I installed the new block. She beat everything that was there."

"Well my Testosterossa here is pure stock with original factory parts. My CHARLIE Frank takes care of her. He has both a Ferrari factory Kno-Pack, and a Do-Pack, so he's good. He can fabricate any new part from archives within an hour usually", Jim brags back and gestures while looking at his CHARLIE Frank.

They both walk over to the front of Steve's rig, and he pops the hood.

"Very impressive indeed!", remarks Jim, "What she develop, about 4?", he asks, referring to thousands of horsepower.

"Bout 5.5. I have a few special mods of my own making in there.", Steve brags with a smug and arrogant look on his face.

Steve was clearly proud of what his car would do, and wanted to challenge Jim and was aching to baste his giant ego in Jim's juice. By this point, Steve was strutting around like the cock of the yard, twitching and bouncing about in a swagger that made Jim Humphries get just a little angry.

Jim thought to himself, "This guy is a real jerk, a real looser, with a hugely inflated ego, and something to prove."

"So, you gonna show me what you got in there or what? Tell you what, I'll go ya for tickeys, what do you say, you got the garlic Jimmy?", Steve taunts Jim with a race for car titles.

Finally, Jim beginning to get irritated and against his better judgement, can no longer stand the taunts, so he went over and pulled his garage door open. He then gestured to Frank to bring his rig out, and the CHARLIE did as he was told.

Jim replies as Frank pulled the rig up to the two men, "I tell you what, how bout I just take you for a ride in mine. You keep your tickey ok, I have a special course."

The rig was unbelievable, and Steve's Jaw dropped as he looked over the beast. It was virtually silent which was unexpected to Steve. It was about 24 foot in length with 2 trucks that tandem pivot in between on 2 planes, both in the horizontal and vertical directions, and 4 wheels each. Each section had its own power plant which were small boxes about 2.5 foot in all 3 directions, but small is deceiving. Jim instructed Frank to install special CCM power plants just the week before, and jim had yet to fully open them up on the course, so he was excited that an opportunity had just presented itself.

"You got to be kidding", proclaimed a dismayed Steve, "That thing? What you gonna do with it?"

"Get in and you'll find out", replied Jim as Frank opened the door for a hesitant Steve.

Both Steve, and Frank got in, with Frank in the back trailing cab, then the rig pulled out.

Jim inquires of Steve, "You hungry at all, have you eaten recently?"

"No I ate something a few hours back. I'm not hungry right now. Thanks for the offer though. Where we off to?", asks Steve, as he stares out the window watching the scenery go by.

Jim grins wryly and replies, "Thats gooood! You'll see, trust me, you'll like it."

They proceeded through the surface streets and while there is usually not much in the way of delay, Jim was not one to let things impede him whenever he is moving and in control. As they approached a cluster of CATs that seemed to slow the flow of traffic, at a distance of about 300 feet, Jim punched the accelerator. A loud whooosh was very audibly heard as the twin HCMCF CarbAunti engines opened up. 90, 130, 160, 195 the speedometer read. The beast got to that speed in 3 seconds, but Jim was holding back.

"There it is, nice!", Steve remarks, referring to the heretofore in-audible engines, as the beast demonstrated unprecedented acceleration at 6 Gs, which laid Steve back in his seat.

While they seemed to approach the cluster of CATs rapidly, and looking through the front windshield, it was obvious that the CAT cluster were quite aware of his approach, as they quickly scrambled to plot and execute a precision trajectory path down all 12 lanes for the impending oncoming missile. The street jogged slightly to the right with an offset of about 10 feet as it proceeded, even so, the CATs were quick; they moved to the right, they moved to the left, and they exited their current path altogether, moving to other avenues; all working together to execute the computed task at hand, which was to move enough volume of CAT, metal, and human occupant flesh out of the way in order to avoid any possibility of collision. It was like the parting of the Red Sea, as Jim's beast cut through at high speed. They moved like a giant swarm of bees, or a fast moving school of herring, or even the way a flock of small birds or bats move while chasing their meal of a swarm of mosquitos, all in unison, one direction or the other, and very gracefully, except that nary a one did even intimate at a bump or scrape along his neighbor, contrary to what one might expect.

Each CAT alerted it's occupants of the situation by an emergency beacon brightly flashing red, along with an annoying ear piercing beep. It was made clear on each CAT traffic nav console screen that the 1500 pounds of metal and flesh were bearing down at an alarming speed. The danger and emergency situation was illustrated by a brilliant red flashing blip, indicating Jim and Steve, and a brightly lit path flashing on the display console street map, which indicated the current trajectory of their approach, and it also illustrated a computed estimate of their likely continued forward path.

On all vehicle nav systems in the vicinity, Jim, Steve, and the beast shown like a fiery red blaze in a backdrop sea of dull blue dots, each one representing a member of the surrounding CATs. Although it clearly informed all involved of the most likely path open for the beast to take, it was only an estimation of that path as it was not possible to know with 100% certainty which path the missile would take, only the most probable, so that is just how the swarm moved to position themselves, all within a few seconds.

As the beast neared the scrambling hoard of CATs, Jim suddenly let go of the steering wheel, but the beast nonetheless maneuvered itself through the path, jogging the 10 feet to the right exactly along the route that had been computed, with all the grace and precision of a machine, at full speed. Jim's beast was fit with a Q-Buddy unit which was an aftermarket autonomous navigation system similar to what was at the heart of the CATs nav systems. Q-Buddys are Quantum Computing Device based control units that give the beast the intelligence needed to compute a probable safe course of travel, and if necessary, to steer it at high speed through the complex maze of swarming CATs, avoiding collision. As expected, the path predicted by the control beacon negotiated the path the Q-Buddy computed, and both agreed, the path was the exact path taken.

As the beast blew past the throng, many of the occupants expressed more than a little indignity through the CATs windows with numerous gestures and choice words for these apparent belligerent and uncouth scofflaws. Shots were even fired. This transpired in quick manner as the next cluster up ahead seemed to have already prepared for the inevitable dispersal, and organized themselves in the same fashion. Observing again through the back window and rear viewport screen, the CATs then scrambled to close the gap and resume their previous intended path.

Jim grinned ear to ear but said nothing. Steve cackles while rocking back and forth in his seat clapping his hands in delight, "Alright then brother, I love to do that too. I do it almost every day myself. I love to watch them things scramble. Its like art!"

This precision action was the result of engineering the autonomous vehicle traffic control systems to handle emergency vehicles and other contingencies such as a vehicle out of control, or those unable to stop, or it seems, even this. It only makes sense that during the planning and implementation of the system, the designers would not have anticipated that some people may find sport in exercising that aspect of the system just for the fun of it, but neither Jim Humphries, nor Steve the dude were the first.

Overhead traffic surveillance systems caught the stunt on camera, so Jim's console immediately indicates he has just received a traffic violation from the old defunct traffic enforcement system.

Steve laughs and says, "Man, you're in trouble now!"

Jim just smiles and ignores the message. He reaches over and deletes it from his cloud-space. Their immediate destination lay up ahead a few miles, where Jim has promised Steve the ride of his life.

CHAPTER SEVENTEEN

Headwaters

A Turd Floats to the Top

[MD Narrative] At some point, the owners, investors, management, and boards of directors of the corporations that employed and benefited tremendously from autonomous technology began to see a trend that was alarming, but the profits earned from the technology were vast, so it was easy to ignore the eventual problems that seemed to be imminent in hopes that they would never materialize, but many knew it was simply a matter of time. My father Boirix was one who knew early on because he was on the forefront of the technology and really understood the potential for significant problems in the business world. Having been concerned early on, but not having the ability to alter the inevitable course, it was not until much later that he began to appreciate his vision and prediction of the societal dimensions, but by then it was too late.

Early on, he covertly set up a working group and council of very smart people to deal with the potential rise of these more potentially devastating contingencies, although answers to the problem were difficult to come by. It was secretive because he had received much of the original budget and contract through the US government, and they did not appreciate their resources being used for covert projects in which they had no control. Soon though, even many in the government began to get a dim picture of the future, and so were more open to Boirix's projects.

That is really when I got personally and morally involved, then witnessed and began to understand much of what I am relating now. I initially worked under Doctor James Bowen, the director of the project as a specialist in the research department, then moving to special projects under Boirix. Since then, and up to now, I have been working to find ways to keep a potential metaphorical rhino stampede from starting.

Hegema Dilemma

[MD Narrative] The advent of the fast approaching paradigm came as no surprise to the minioned echelons. Although not knowing the exact form of the era, they nevertheless saw it approach, and may even have had a hand in ushering it. Having acquired vast resources, they were in a position to know and act upon virtually everything, so they welcomed the approaching era. However, their vision was limited, so they were unable to see the landscape past the transition, therefore, they could not see their own fate. Although, the minority amongst their ranks were to view the approaching era with caution, the vast majority of them believed headlong that the new paradigm was ordained for them exclusively, and they alone would derive the benefit, while the inevitable fate of the balance of humanity would unfold in passing. They believed the coming age was not within their power to shape or alter to any appreciable degree, therefore they were not culpable, regardless of deed; it was simple recognition of the inevitable coming age, to fulfill the destiny of humanity as foretold. They believed they were not as the rest, but specially endowed, as those alone destined for glory and lavish splendor, apart from those destined for oblivion. With brazen circular ethos, their beliefs shaped their attitude about the coming age, and their attitude justified their belief, their action, and their plans for murdering the world.

§

Deep in the formidable bowels of Hegemony, Wyoming, the business of running the affairs of the planet continue throughout the second day of the regular session, and late day there is a call for a full panel, and a full house of attendees. The great agenda has been determined for the session, a high level indeed, near crisis proportion issue emerging, needing consideration and the full attention of all.

The question hangs in the air like a bad stench. It weighs heavy amongst the rank and file of the elitists, having been first noticed and brought to light by those involved in commerce and industry; now it concerns their high counsel. It was brought to their attention earlier and designated to investigation, then elevated to priority. The observed phenomenon that gave rise to the issue were assigned to several of the usual minion army of analysts in the myriad of think-tanks, and policy institutions, who have done their analysis and compiled a voluminous report with the usual tedious blather of useless dry facts, figures, charts, and statistics, designed to obscure the essence of the matter, with little in the way of practical interpretation. Now it has been delivered for consideration and directive to the panel and the session, who have the power and the keys for cutting through the purposed noise in the language, and the power to set in motion, whatever course of action is necessary.

Cutting through the swirling collage of misdirecting nonsense, and focusing on the essence of concern, the simple question has emerged, "Why have global profits started to diminish in the face of record levels of global productivity?"

At first glance, the problem seems innocuous enough, if just a temporary anomaly, but some have understood the magnitude and fallout if the trend advances, and have voiced concern. The analyst's consensus seem to be that there is nothing to be alarmed at, it is an inevitable unavoidable occurrence; to be expected; from the hegemonist point of view, it should be considered all part of the plan. The question amongst a minority view then becomes, how should it be handled to make sure it does not become a crack in the dam that holds back the raw weight of civilization, so as to bear down directly upon them. Power is powerful, and will destroy those that wield it, if not done so properly.

Before the panel and the full assembly, the advocates for both sides are speaking in place performing for their audience. The first to speak, a man with the organization's given name of BelShazzar opens the discussion with a rousing appeal for how far they have come, and what is at stake if the situation should get out of hand.

He also makes a case that it is an inevitable course and must be endured, while transitioning to a new era, "The situation has been studied, and analyzed, so we are very apprised of the implications of the approaching paradigm. It is the foreshadowing of what we have believed would come and has been foretold for eons. It is the fulfillment..., the absolute apex of the promise of science. In some ways it represents a great deal of our aspirations throughout millennia. It represents a time when we can throw off the shackles that now bind us, and the time when we will finally receive the inheritance promised us. A time when the parasites that now infect our destiny, are finally cast off, a time when we shall trod upon their ashes."

He continues, "We shall not enter into that paradigm, unless we pass through the events set to unfold before us, that even now show definite signs, and stirrings! This is a time to rejoice, 'for our redemption draweth nigh!'"

The counter advocate, dubbed Xerxes then speaks to answer the assertions made, "The oracles and seers that have looked to the time, and, if indeed we are at the advent, we are blessed and privileged to have witnessed the arrival. It is indeed blessed, but have they, the oracles, seers, and sages, of the ages, really peered through the haze and the fog, to clearly see the other side? Are we so assured that the other side favors us? Or could our vision and understanding of the time simply be self delusion; a myth; one that both you and I have been told repeatedly, which has misled down the primrose path to shape our confidence? No, I am afraid what we have all believed about the clarity of the coming age, is nothing more than wishful thinking, not reality. It is the myth that concerns me, and should concern every one of you. Much wishful thinking has been espoused by dreamers, but the reality is, we stand at the threshold of unknown order, and there is no real assurance or knowledge of how it falls."

Continuing, "Now, having stated the need to separate fact from myth, I must tell you that personally, it is my conviction that we will emerge firmly in control, just as we have been throughout the ages, as I can see no reason for things not to continue as before, but we must examine the risks. I would advocate caution. We must oversee and shepherd the process to ensure it proceeds in a way that may not leave our hands. Clearly there are unknowns, so we must mitigate the risks, and the possibilities. Perhaps my opponent can present exactly why we should give the issue no mind, and rest assured that things will naturally, and unquestionably go our way, without a hitch. Keep in mind, we wield a great deal of power now, but I shudder to think of the consequences, should that power get away from us!"

The other man, BelShazzar perks up, and says, "I would be glad to make my case for the certainty of our complete delivery into the future, just as has been foretold. The future belongs to us. It has been put forward as a cornerstone of our credo for millennia, and, although, I would like to say there is no doubt, clearly there are misgivings, or perhaps, misconceptions; doubts within our own ranks; doubts that need to be laid to rest, to minimize confusion, and bring absolute clarity. We can not afford to proceed with division amongst our ranks."

BelShazzar continues, "It is assured that the vastness of goy, in the dullness of their minds, may get a glimpse of what lays ahead, but when has that ever helped them, or denied us? With what initiative have they ever successfully leveraged knowledge that might better their lives, much less save them? They have never used anything to their advantage against us, which is why we have come to this very time and place. It is why we are in control of the now, and why we are in control of the future. I, just as all of you, have ample reason to fear a world in which they are free from their prison of order; free to order the world as they would, denying us our right, but again I ask you, when, in the last several millennia, have we seen their time? We have always seized control as far back as history records. So as in the past, so it is for the future, we have nothing to fear."

Considerable murmuring and furor stirs amongst those present when suddenly a man interrupts the proceedings, breaking protocol. He stands and speaks out of order, then several others raise their voices chiming in, in agreement, which demonstrates palpable fear amongst the ranks of the minioned, who have always considered themselves above panic or apprehension. Weakness, and vulnerability are shunned.

One asks, "They are rioting now and have been for weeks, and they are gaining strength. They call themselves belli-warriors, and fashion themselves after the man, Belisle. They protest the machines. Are they among us or not? Does he do our work or is he off the reservation?"

Another one asks, "If our profits continue to shrink, how do we maintain control over anything? You are lying to us. This is not a good situation for any of us!"

Still another one stands and rails loudly, "Many have already left our company as if they could care less. If they believe they don't need jobs, they can not be controlled, and we will be at their mercy, what can be done?"

The Mist Driving a Storm

There are times in history when Political abuses and scandal are muted and quietly swept under the rug, and other times when crimes of much lesser magnitude or significance, but when discovered, can lead to massive change or revolution, with a very powerful and catalyzing effect on a desperate and angry population. This was a transitional period with assorted forces at work pushing and pulling at the fabric of societal order, and when the tear finally occurred, it was massive in scope, unleashing untold power, but was also liberating and led to a considerable renewal of civilization.

Circa 2055, the unemployment situation in the US and other countries was mounting, mostly because the productive capacity of the manufacturers in the US and other countries had begun to accelerate. Both the previous generation of Dalbots units, the ARCH6s' and now the much improved next generation CHARLIEs', along with several other manufacture's labor replacement hardware, were becoming ubiquitous in many previously held blue collar and unskilled labor jobs, the Dalbots units having by far the greatest, and highest profile presence. Employees of these companies were being given instruction on the use of these machines to aid them in their jobs. They were asked to learn the basics of initializing the machines and then employing them on their rounds all the while suspecting that they were in reality, metaphorically, digging their own graves. These companies began to experience considerable gains in productivity and by extension, profitability by this, so much so, that it made it very easy to make 'work force reduction adjustments', which is simply to say, they fired massive numbers of these people, and replaced them with their mechanical counterparts. Needless to say, this did not sit well with the rather loud and spiteful humans being replaced.

There was a great upsurge in the unemployed masses, created by wholesale displacement of their jobs by machines. These people, both American and foreign born, white, black, latino, asian or mid-eastern, all are in the same boat, and unhappy because they are worried about how they are going to ever maintain or build their lives if they continue to see their jobs replaced by technology. Contrasting this, is the record windfall profits that the corporations are now raking in because of the massive advances in productivity. A product that used to cost a dollar to make, now costs a dime, but the marketplace price has only dropped by a quarter, with the difference accruing to the bottom line, and this whole trend accelerating.

§

Mid-morning, and five large 50 ton container-dump trucks rumbled down the avenues toward their target, a large and well traveled intersection in downtown Milwaukee. The 5 seemed out of place amid the throng of CATs that are the usual transport in that part of the city. Even though they were on different streets and coming from different directions, they were obviously converging on their target simultaneously. Whats more, is that they were manned by several persons actually at the auxiliary emergency wheel, driving them, or at the very least, they were instructing the truck's pilots where to go and turn. The planners had timed and choreographed the operation before-hand through simulation, and the event seemed to be going off without a hitch. The first of them arrived at the large crowded intersection in downtown Milwaukee, and had started setting up a barrier by dumping the contents of 4 of the large trucks strategically blocking the 4 avenues of the intersection, then the operator of the 5th truck, a flat base rig with a large bulldozer, unloaded the large machine right in the intersection and it set about moving the various mounds of debris around the area blocking it off from the approaching CATs that would not attempt to pass through such a congested area, so they just sat until the backed-up jam of CATs behind the barriers reversed coarse and eventually escaped the clogiture.

The bulldozer eventually had succeeded in blocking all of the avenues of ingress, and left a large pile of construction debris in the center, which he then parked the large machine next to, leaving a corridor about 15 feet across in between the pile and the machine. The clogiture of CATs backed-up behind the barriers, eventually cleared and people were once again well on their merry way, but not before it caught the nose of cloud-cast reporters; a story was getting underway, and building. Within hours, several blocks of buildings surrounding the vicinity were packed full of curious gawkers in a carnival atmosphere, many preparing for the long stay with lawn chairs and sleeping and camping gear; and the usual contingent of police, and special forces; and the myriad of alphabet agencies swarming around setting up structures, and attempting to assess the nature, and threat level of the hubbub. This kind of event is taylor made for the attention seeking government agencies, the scurrilous media, and the wanton bloodlust of the public alike.

The Autonomous Economy I - The Strong Man

Belisle Whetstone leaned back against the assortment of crates, tires, chunks of concrete and road asphalt, and various other forms of debris that were erected in a formation reminiscent of a rampart used to construct a medieval siege-works. A portion of his small group of devoted followers were there, mostly the true-believer types that were dedicated to making some kind of change. They followed Belisle and believed in his vision explicitly, and all swore their allegiance to him and his cause, even to the death if necessary, which was not really a big thing because there was virtually no chance of that ever really being put to a test. They were mostly rabble from the underclass and the street, disaffected losers and people that were generally looking for something to give their otherwise pathetic and mundane existence some meaning. They represent a brewing storm, and Belisle was the mist driving them.

Belisle Whetstone had grown up in significant opulence, but during his formative and college years, he began to take on the all too familiar characteristic syndrome of those that inherit wealth. He felt insignificant and trapped by what he observed by his families position and stature, and did not want to be owned by the material trappings and positions that he had not chosen for himself. So instead of gracefully declining the wealth that could have been his for the taking, or choosing to live a humble life harmonizing with the blessings of his family, he instead chose to burn those bridges and became a follower of many in the anarchist movement.

A mug-shot image of Belisle is emblazoned on the 15 foot screen in the large office on the 3rd floor of the Wisconsin office of the FBI, '*The Federal Bureau for Instigation*', in Milwaukee which sits very close to where Belisle's group is staging their event.

Special Agent Brian Girardano is speaking to a small group of agents and officers, "He advanced quickly as leader of a violent fringe faction linked to the mainstream groups that had begun to gain significant stature in the last several years. He was recently jailed short term for an attempted bombing of a building owned by a company, of which, his father sat on the board of directors. Normally this kind of act would have carried a pretty stiff penalty; decades in prison; but his family intervened and his defense successfully pinned the whole thing on his partner in crime, Randy Vogle."

Agent Girardano clicks on the pen and changes the picture to that of brothers Randy, and Jake Vogle, then continues, "…who in reality was just one of his foot soldiers, carrying out his orders. He's an ace, a real donut hole."

Belisle and his crew are inside of the makeshift tent, made by stretching covers over the corridor between the debris and the bulldozer. Jake asks Belisle. "I'm with you garl, but tell me again Bell, what is this supposed to accomplish, we have tried this kind of stuff before and my brother is sitting in prison right now, because he believed in you, and you let him take the fall. I'm not so sure that I want to follow him there after all?"

He replies, "Relax garl, I told you, your brother wanted me to save myself for the cause, because I'm vital man. He sacrificed himself, regardless of what he told you in retrospect. It's going to work this time because we have covered everything; they are weak and really have no excuses. The nation is with us this time, I really sense it; besides, we are already up past our necks, no turning back. The worst thing is a citation for destruction of public property, and disturbing the peace."

Belisle addresses the entire group who are looking somewhat scared and unsure of themselves, and seem to have doubts about their fearless leader. As he is doing this, he proceeds with putting on the uniform of the construction company that employed the CHARLIEs' in his possession, which consists of a vest with brightly colored orange striped inserts, and a yellow hard hat, and the letters BRC emblazoned on the back.

Belisle approaches one of the CHARLIE units which snaps to Belisle's attention. He then demonstrates how to torture the CHARLIE by grabbing it aggressively and talking through the procedure, "Grab it around the mid with one hand and wrench his arm back like this, to a point that makes it scream, like this!"

Belisle demonstrates his technique on a live CHARLIE that was part of a compliment of machines recently commandeered off the street by his crew from a nearby construction project.

The machine protests, "Sir you are hurting me, please stop."

Belisle then hits the machine with a piece of pipe about the size of a baseball bat, which knocks it to the ground. The CHARLIE protests again, "Sir that was not necessary, please state what you would like me to do and I will comply."

Belisle then picks up a large cutting or slashing weapon attached to a long pole resembling what would have been described in the medieval age as a glaive. He then places his foot atop the prone CHARLIE and proceeds to insert the weapon between some of the protective plating on the torso, then to pry the plating apart, "I..., would..., like..., you..., to..., scream.", he replies with heavy breathing timed to every lunge.

The CHARLIE then complies "Please help me, aaaaahhhh, please sir you are hurting me, aaaaahhhh, aaaaahhhh."

"Louder..., louder", demands Belisle, "Louder..., and cry more, you piece of shit."

The CHARLIE continues to acquiesce to Belisle's demands, "Please stop aa-aah-haa-aah-haa-aah, aa-aah-haa-aah-haa-aah, please stop sir, please stop."

Belisle then stops torturing the CHARLIE and says to the rest of the group, "Thats how its done, any questions? Jake, you take the lead on this will you, and Brenda, you're with me", as he gestures to the two eager participants gathered in the small group.

Belisle then loudly and excitedly continues bolstering up their courage with a rah-rah speech, complete with the gusto of a high school cheer leader squad, "Are you with me? Do you have the rocks it takes to stand up and be counted? Do you want to be remembered for having the guts to take a stand when no one else would?"

To each of these he gets the obligatory affirmative cheer of, "yea!", and "Damn right!", but there is still an uneasy chill amongst the group who proceed with their planned stunt.

They were there to demand they get their way with usual tired variation on old pieces of leftist drivel, such as democracy and equality and social justice for all people. They were convinced of the correctness of their demands based on the fact that they felt good about it in their hearts, and because if their motives were pure and self-estimated as selfless, then that constitutes truth, and justice was implied. They had never considered the fact that if they were calling for democracy and an equal voice for all, then they had failed because they demanded that everyone agree to their rankled point of view, and that the nation should allow them to be the architects of all social order. What self aggrandized hypocrites they are, with their eyes fixed firmly on their short sighted narrow little cause, without regard of the opinions or rights of others in the democracy they are calling for. Nevertheless, they are getting attention because the cloud-caster networks from around the globe are there and many eyes are upon them as they pull their choreographed and scripted stunt, cleverly designed to prey upon the vulnerable sensibilities of a gullible, shallow, confused but somewhat sympathetic nation.

The group threw back the front tarp section of the large tent that had been covering their position from the prying eyes of the cloud-cast networks, they then walked out in the open but still surrounded intersection that was now a ringed fortress. Belisle led the way followed by Jake who was towing two of the participants, who's faces were covered, and who were tethered by a rope. Their hands were bound, and they were tied together back-to-back, as if they were hostages put on display. At this point, none of the police or other law enforcement authorities had dared breach the perimeter established by the debris because they had not yet assessed the situation. The two people in tow were then gestured by Jake to sit on-top of a large pile of concrete debris, positioned strategically for the purpose of making sure they were in plane view from afar. Jake then positioned himself behind them about 5 feet and sat down holding the rope they were tethered to. It was plane to law enforcement and the cloud media, that these people were innocent pawns used by a sick and sadistic gang of dangerous terrorists.

Belisle then emerged from the debris corridor with a bull-horn in his hand. He walked over near where Jake and the two hostages were seated and began to deliver a speech through the device. He directed his rant to where he knew the media had staged itself, giving law enforcement authorities a clue that he was seeking media attention which merely affirmed what the FBI had already determined.

He rants, "For too long, we have stood by while the global corporate world has gained record profit at the expense of the common worker, they have continued unfettered, systematically gutting the human labor force. They continue to feed like vampires on the blood of the common man; we are not chattel; we are not slaves; and we would not accept that in principle nor practice; we are human beings not animals, not toys, not a line on a balance sheet. The people will standup to take back what is theirs!"

On and on, Belisle barked his manifesto for hours while law

enforcement waited and watched, then finally attempted to make contact with the group. Although, he humored them to delay, Belisle was not interested in talking with the police, only the media. He then set his bullhorn down and made a gesture to one in his group to bring out the CHARLIE he had used in the demonstration earlier. The young man walked the machine out to Belisle, and then went over to relieve Jake, who had sat holding the rope and listening to Belisle rant. Jake then got up and went over and put on the vest and hat which Belisle had worn earlier, then took hold of the CHARLIE and began to bend it over and wrench his arm back as Belisle had shown.

Jake then apologized to the CHARLIE, "I'm sorry that I have to hurt you now, its not personal but I want you to scream and cry loudly."

The CHARLIE complied, "Please sir stop, you are hurting me aaaaahhhh, please sir you are hurting me, aaaaahhhh, aaaaahhhh."

Each time Jake applied pressure on the limb, the machine let out another groan. After a period of warmup, Jake grew more confident and felt less guilty about really putting the screws to the CHARLIE, so much that the machine began to roll its eyes back in the sockets, and shutter them rapidly while repeatedly expanding and shrinking the apertures of the eyes, or shrinking one while simultaneously expanding the other rapidly. This was done with great effect while the cloud-caster's autonomous intelligent Hawkeye cameras were able to zoom to extreme closeups of the spectacle, even from a significant distance, which went out live, and was extremely shocking to those watching.

Belisle then grabbed the bullhorn and moved it close to the crying CHARLIE so as to maximize the effect the media would induce in the public by showing the spectacle. The act was in part calculated to gain the attention of the public for what a cruel act would induce if it was perpetrated on a living being, the public being naive about the facts of robot pain, and in part to send a message, a sort of wake up call to the world, or at the very least, certainly the greater Milwaukee area. Belisle reasoned that the screams of the CHARLIE would elicit outrage and sympathy in the minds of those that viewed the event even though it had previously been widely disseminated by the media that the machines were not really able to experience pain, nonetheless, Belisle wagered that there would be a large sympathetic response, and this was designed to focus all eyes, and attention on the man and his message.

He had worked previously with other commandeered machines in anarchist workshops, with the result that even though they did not recognize Belisle or his comrades as masters or owners, they would to a limited extent, nevertheless, obey commands issued by humans they did not know who were in charge of them by virtue of the fact that they were captive by the same, and were programmed to comply with people that wore the construction company uniform. By this time, the fax-sentience feature had matured through contact with, and the experiential exposure that comes from millions of units interacting daily with their trainers, and other human experiences to an extent that these machines could display a convincing expression of pain, torture, and agony for the masses.

Ironically, the tactics Belisle employed were antithetic to the message that he wanted to convey. He solicited sympathy for the machines as he tortured them, but he wished to convey at the same time, the message that the machines were really a menace, and as tools in the hands of the global corporate cabal that employed them, in reality, they were the cause of real anguish and misery amongst the human population. Whatever the logic, no matter how twisted it may have been, it accomplished that objective.

In the past, when similar provocative events were carried out by radical leftists, and seditious-incendiary malcontents, and which usually covered the media flow to the point of saturation, it was known that the likes of the Belisle Whetstones' of the world were not alone in their charade, but sanctioned and protected tools, and were usually enabled by legions of facilitators with access to large amounts of money, resourceful planners, backers, scripters, directors, facilitators, agitators, and the contingent host of paid actors, employed mostly from amongst the ranks of the elitists and the true-believer minions of the globalists cabal; who staged the events for the consumption of the ignorant manipulated public. Their stated objective, a fait-accompli declaration that, *"We hold the hegemony of the world in our hands!"*, still presently characterized only as a working motto toward the day that they could openly and proudly affirm it to the world.

This event was not sanctioned by the elitist cabal, so Belisle committed a sin against his avowed lieges', by acting out of place. Nonetheless, it had been done, and nary a word nor deed was uttered by them, more likely confusion and curiosity set in by these potentates, a kind of consensus resolution of, *"This is much like our work without our hand. We'll watch to see what transpires."*

The stunt that Belisle and his followers had staged carried on for over 16 hours then ended, because fatigue and a sense of not really knowing what the point was, or, as some wondered, if at the risk of their freedom, they hadn't just blundered into a mistake of monumental proportion that accomplished nothing but the entertainment for a few hours of the masses of the idle and frustrated. The truth is, some of these people admired Belisle and his motley band of boobs because so many of them sympathized with the muddled message which Belisle had tried to convey, but at the same time, some felt sympathetic to the machines that could feel pain, experience grief, and were apparently more like humans than the public had been led to believe.

Although local Milwaukee police were first to be called, and first to respond, the local office of the FBI being within a mile or two of the event, gave them the impetus to move in and take control of the situation. Although normal protocol dictated they move their Mobile Command Center (War Wagon), equipment, and men, in as quickly as possible to contain the situation, Special agent Brian Girardano was guarded and cautious. He had moved in with men and equipment prudently and was on scene for the most critical window of time, which was within 4 hours of their first report.

Agent Girardano was perplexed because no one from the higher food chain echelons had advised him that such an exercise or staged event was slated, which was the normal protocol when this type of thing occurred in any field office's backyard; usually someone knew what it was about and data poured forth. This time there was nothing, so he put in a call, but had not heard anything back. He had held back not wanting to escalate the situation, while not knowing what they were up against other than what the local police had already determined which was the names of two of those involved, Belisle Whetstone and Jake Vogel. He also just needed time to get his people and gear in place while attempting to assess the situation. He had agents with high power scoped rifles set up at elevated positions in a few buildings surrounding the intersection square awaiting instructions.   He also employed a contingent of 3 CHARLIEs' equipped with special communications gear and Hawkeye cameras.

The Hawkeye cameras were developed in response to the need of industries that require the audio/video capture of all aspects of any random event.   They could be strategically positioned around the perimeter of an event such as a sporting event or war zone, or newsworthy story, and because they were networked, and could communicate with each other over the distances involved, they could coordinate and prioritize the audio/video capture of the details of any event that may transpire.   These cameras have up to six lenses, each with very high resolution capture elements, and employ the artificial intelligence of a bird of prey, thus the name Hawkeye.   They can prioritize and coordinate their view on the event, with up to 36 different observations from up to 6 different vantage points, each with 6 lenses, and are programmed to immediately zero in, zoom in, and track any motion.   Some are mounted on CHARLIE units, and some on autonomous vehicles.   If something moves, the camera will immediately locate and zoom in, focusing at a great distance within a few milliseconds, then track that motion.   The mobile units can then track a moving event over distances.   This way they can get an extremely detailed view of each and every object in motion, such as in a football game play when the ball is snapped and scrimmage is played, every player of significance is captured by several cameras from different vantage points.

As with any media involved act of propaganda, when the simpleton masses of the naive and credulous are engaged, it results in considerable confounding of the reality of the matter.   In many respects, not just the most obvious one of the errant notion of machines sentience, believing them to be on the same cognizance and emotional level as humans, but also in their actual role as laborers in society.   Although, from a confused message that Belisle had disseminated, the latter was much closer to the truth of the matter than he realized.   Some of the more colorful observers feigned an air of rage, which at that time was popularly dubbed 'stage rage', nevertheless it had tremendous impact and spawned several genuine movements.   Some of the less enlightened, demanding laws that would protect the civil rights of these machines, with other geniuses demanding the convening of a global commission to look into labor and pay issues, to make sure there was equity and not any exploitation of the machines, stating, "Its obvious they are not treating the human workers very well; we should stand in solidarity with the robots, to avoid them being exploited." Despite all of the idiocy that ensued, out of all the insanity, there was a genuine movement of purpose, although violent, and somewhat blind.

The FBI presence onsite sent a CHARLIE toward where the group was hunkered down, which was fitted with a Hawkeye camera that gave a remote view of the ground level at a closer range with in alternate light spectrum, allowing agents to observe things they would otherwise not have a close view of, such as bodies that were not viewable otherwise, inorder to better assess the situation. The CHARLIE was able to determine the total number of people and captured CHARLIEs' kept in the makeshift tent setup, as well as a closer view of the two individuals who were ostensibly being held hostage. The CHARLIE gave agent Girardano the ability to call out to Belisle at a closer range in attempt to get him to talk to the FBI, which Belisle would only do while attempting to buy time to allow the maximum effect of media coverage and its apparatchik effect on the population.

Agent Girardano walks into his mobile command center on location, having just conducted a walking survey of the perimeter of the theatre of interest. On the way in he stops to check that his com device is working properly, because he has still not yet heard anything back after sending a desperate request for briefing and data from on-high regarding the event in his backyard, with which he is tasked with resolving. He begins to believe that apparently this is not a normal staged and sanctioned event.

Getting the attention of the several agents present, he begins talking with the group, "I have yet to hear back from my superiors; they have not confirmed sanctioning of the current theater and situation, so have not advised me, therefore, we must assume it is real, and live, and proceed as such... Belisle Whetstone is no stranger to us. Many of you may be familiar with the Reichert building incident in Lansing 5 years ago? He and Randy Vogle attempted to disrupt sensitive business negotiations that were taking place then by planting a bomb on the floor just below the location where the meeting was to take place. Luckily, through good investigative work and a tip, the plot was uncovered and stopped before it commenced. However, Belisle got off the hook through the intervention of his family. Agent Blake, can you please give us an update on the situation; how many are there, and what do we know about their demands?"

Agent Blake then reports, "As near as we have been able to determine, there are 6 people in theatre including the 2 tied together sitting out in front. They appear to be hostages. The leader of the group, as you've assessed, appears to be Belisle Whetstone, and Jake Vogle is there as well. We have not identified the others yet, nor the hostages, uh, mainly because we can not get intel on their obscured faces, only their body signatures, which also applies to the hostages, but we are working on it and expect to know who they are as soon as possible, uh, as near as we can tell, they have misappropriated 3 to 4 of the Dalbots CHARLIEs from Blanch & Ruud construction which has a large project going on in Madison."

Agent Girardano interrupts, "Stop there Agent Blake, uh, you know who owns the robots? Have you contacted the owners, to have the units remotely shut-down, or how about having the local authorities override the protocols and call the units for emergency service? They should have been briefed on the emergency service protocols by now I would expect, would you not agree?"

Agent Blake continues, "I have personally been in contact with a mister Jensen, one of the owners of Blanch & Ruud Construction."

As he is speaking, agent Blake is flipping through an electronic tablet device looking for the contact information and the recording of a telepresence session with Mr. Jensen, then he continues, "I spoke to him at one PM this afternoon. The company reported the units missing six days ago to the police in Madison, then they exercised the homing location and remote report protocols, but the machines have not uttered so much as a blip back, which would indicate the units have been compromised. The only way we know for sure who the machines belong to, is that we identified the uniform worn by one of the perpetrators, mister Vogel, and surmised the machines belonged to the same company..., and because there are not too many anarchist groups that can purchase eleven million dollar units and then abuse them. I subsequently made contact with mister Jensen who confirmed four units missing. As far as the Madison or the local police invoking emergency protocols in an attempt to solicit a response from the machines, well, both are equally useless."

Agent Girardano then asks agent Coleen Yiu, "Agent Yiu, has your group been in contact with the machine's manufacturer to solicit their tactical help in understanding if and how the machine's security protocols may have been breached? We need to understand which security vectors may have been compromised to where the units will not, or can not communicate with their owners."

Agent Yiu then responds, "Sir, we have made contact with Dalbots, and spoke to a mister L Mahendran, who is a security specialist with the company and explained to him what I could of the situation. They do not believe that a security breach has taken place, because the security P&D, or protocols and directives which they call it, adhere to very high, national security high level standards. Mister Mahendran suggested the communication system may have been disabled in some way..., or there is something blocking the radio signal, preventing communication. Thats all they have said except that they will stand-by and offer whatever help they can. They even offered to fly one of their field team out to assist us if we ask. Shall I ask them to send someone sir?"

Agent Girardano then replies, "What do you believe Agent Yiu? No, not as of yet Yiu, we need to determine the exact nature of the threat before we involve others."

Agent Girardano continues, "I did some fieldwork, as part of a research team on low level domestic terrorism a few years ago, and the name of Belisle Whetstone came up. He received advanced degrees in political and information science when he attended university some years back. He then put those degrees to work as a regular instructor at some of the anarchists workshops held at various times throughout most of the last decade up to now. We believe that..., uh..., Agent Yiu make a note. I want you to get back to your contact or someone at Dalbots and see if they have any record of mister Whetstone having ever worked there."

Agent Yiu then replies, "Yes sir, immediately!"

Agent Girardano continues again, "The case was made a minor case study because at the time, there were no collateral..., uh, casualties, but the psychological and other profile characteristics lent themselves. I mean classic, textbook, the analysis predicted something like this was likely future for this guy Whetstone. The others are textbook followers, driven by the need for acceptance and validation, and Belisle Whetstone is the classic pied piper figure. I went back this morning and reviewed the file. The investigation concluded that Whetstone is very adept at security systems penetration, and reverse engineering others. He was able to penetrate, and circumvent the security of the Reichert building. He has since been suspected of consulting on the acquisition and modification of advanced weapons systems, also, no surprise, in the report, it was suspected but never confirmed that the group he was involved with were stealing and reverse engineering these Dalbots units. Several of them disappeared from the greater Milwaukee and surrounding areas, and were never recovered. This guy is dangerous. As always, none of this leave the room. Good work Agents!"

After about 17 hours into the stunt, Belisle and his minions started to run out of steam and began to signal that they had made their point, and were now willing to comply with the commands the FBI issued. All of the participant followers of Belisle came out of the tent with their hands held high and gathered together in a straight line. To the surprise of everyone observing, the 2 people that most believed were hostages, also untied their bonds and joined the others in a prone position signaling that they were in reality, not hostages, but part of the group. The reason that the 2 were there feigning the status of hostages held against their will, was to stop any fast and violent police action designed to end the demonstration. It was calculated by Belisle's group that the people tied-up and sitting on a rock in plain view, ostensibly against their will, would be seen by the authorities as maybe provocative, and as requiring caution on the part of law enforcement, but in reality it certainly would not rise to the level of criminality, after all, there was no one held against their will, neither was there any one in danger of harm nor threatened. This was the genius of Belisle Whetstone who could make a situation appear to be completely different than the reality, while sending a tremendous, and sympathetic public relations message in the process.

The members of the group all came out and stood inline with the others awaiting the move of the authorities to take them into custody which was done quickly and without incident. It was also calculated that the stunt, although disrupting business on a Friday in the busy downtown section of a major city, really did not constitute a threat to anyone, and probably would not rise to the level of criminality. It was a media stunt, nothing more. The courts however, may see it differently. Regardless, Belisle believed that he would suffer no penalty greater than a slap on the wrist and maybe a short stint in jail before his lawyer would get him and the rest of his group out. He also knew that he had made a very loud and clear point that with any luck would resonate with a certain segment of the population, emboldened to stand up with him and send even a louder message. His calculation, very soon thereafter proved to be true, way beyond his wildest imagination.

CHAPTER EIGHTEEN

Effluence

Swine

Subsequent follow up reports on the Philly-Philanthropist, Julian Kennedy Jr. are aired every year for about 6 years right around the holiday. A barely changed Berry Gonzales, who originally reported on the story 3 years before, is standing with Mr. Kennedy on the street getting ready to make a live report. The crowd has built, making around 150 individual groups, totaling about 600 individuals there to see and plead with Mr. Kennedy. They form a large line stretching around the buildings that outline the block for about 100 yards, with a yellow tape barrier erected between the crowd and the streets they line, and are patrolled by a large police presence stationed along the perimeter.

On this particular event, Mr. Kennedy is sitting on a fancy ornate throne encrusted with what appear to be precious or semi-precious stones, and gilded with gold and intricate colored inlay of exotic woods and coral, in his usual spot in the heart of the downtown business and shopping district. His 4 assistants are busy milling about, including the mouth with the megaphone spewing the usual litany of instructions about orderly behavior. Mr. Kennedy had also recently been spotted nearby a leisure and shopping area the week before. Conspicuously, the throne of his queen was also present, but without the queen. Nobody seems to know what the throne is supposed to represent, so much banter is exchanged as the segment airs at Benny's on the old cloud-set sitting on a counter just above and overlooking the mirrored bar.

The Autonomous Economy I - The Strong Man

Berry Gonzales clips his microphone onto his collar and straitens his tie, then starts to speak, "We are live here with Julian Kennedy. You may recall that the last couple of years, mister Kennedy has played the role of philanthropist and guardian angel giving out money and help to people in need, and I must say, this year he is in high style. Some people have referred to you as Santa-Clause. Tell us, do you accept that title for yourself? Are you the Philly Santa?"

Benny's, is a local neighborhood bar, and regular mid-afternoon bar-flower patrons watching the segment are overheard joking and speculating that the man believes he is a King Midas, or King Santa Clause.

The establishment is dimly lit, small, and is billowing in thick cigarette smoke with the stench of sticky and rancid old beer and other booze which covers the floor, the stools, booths and virtually everywhere else, having need to be completely hosed down at high pressure and scrubbed with suds for the last 25 years. It makes the patrons shoes stick to the floor as they walk through. Andy Clausen and his 2 usual barflower colleagues are present with him. All 3 are regulars who seem to have decided to make Benny's their preferred second home. The 3 are barflies busy buzzing about, up and down, into and out of the toilet and at the grungy pool table; a continuous deluge of beer, cigarettes, and chaser shots flowing through the 3 of them.

At about 4:30 in the afternoon, the Julian Kennedy event is just beginning media coverage and Andy is watching with his buddies. Andy then announces that he needs to go and manage the cash in his stipend account at the bank down the block so he excuses himself and steps out the side door of Benny's, and lo and behold, he is in the line for the Philly-Santa only about 6 to 10 groups back from the head. Some people showed up the day before and started to camp out right at that very spot, which Andy observed as he came earlier in the day; but, it had not dawned on him who or what it was about until now. Despite some rather stern words and tussling from some in the line behind where Andy stepped in, he held his ground and was not about to give up the opportunity that now lay before him.

Earlier coverage that day had shown a motley crew of hopeful needy recipients starting to line up, some with severe handicap from birth and some having been injured, and for whatever reason, were not able to get the medical attention they needed. For the most part, these were the downtrodden and truly needy people of various ethnic backgrounds, ages, size and shapes. Some with crutches, wheel chairs, and other equipment, and bandages or braces that looked otherwise normal, some of which may have been faked, and some obviously helpless with grotesque deformities. Later the crowd started to swell with what could only be described as criminals, hustlers, ex-cons, and drug addicts, many of them rough looking, intimidating, or outright dangerous.

Andy Clausen certainly fit in with this later group, and was just about as needy, which is to say, not needy at all. Andy Clausen, along with the rest of the villainous class attempting to blend in would better be described as 'wanty', or 'thievey', but certainly not needy. They simply reasoned that if this guy was chump enough to offer free money to needy people for the asking, they would certainly qualify. In his mind after-all, Andy Clausen's father left him, his mother, and siblings when he was quite young, which made him resentful and he also grew up in a somewhat run down part of town and never got the chance to go to college as some of his acquaintances had. He had also been introduced to the inside of a few jail cells from time to time in his younger days and was not able to support the 2 children from 2 of his 3 failed marriages. In general, in his mind, life had handed Andy Clausen a shit sandwich and that qualified him as 'needy.'

As he approached the front of the line, he could see a rather brutish and swarthy looking man up 2 to 3 ahead of him. The man was asked by the assistant to come forward and stand on a mark, then stare into a camera that scanned his face. The camera and other equipment were operated by the local police who were using it to identify the people that stood in the line to see if any of them had a rap-sheet. As it turned out, being that this was somewhat of a festive occasion of genuine celebration, and not wanting to spoil or taint it with a rancorous scene, Mr. Kennedy's aids had asked the police, if possible, to use the facial identification technology to screen out the fakers, and hustlers, but not to attempt to apprehend any of the miscreants that may show up unwelcome or wanted by the police. In this case, the officer recognized the man had outstanding warrants on several unlawful charges and just asked him to confirm his name. The man replied that it was indeed his name, at which time, the officer said that he did not qualify as a needy person and to please step out of line. The man did so, and proceeded to walk away slowly at first, then gradually accelerate his pace.

The officers that told him to leave were now observing and following him by remotely viewing him at every step as he moved. They observe a map of the area on their cloud monitors showing the blocks and streets along with the location of every CAT as it moves, and every building he passes. The particular cloud-view also shows them the location of all the available surveillance resources from the free CHARLIE units on the block the man is walking on, so they engage 2 of these surveillance resources that display what the CHARLIEs' see as the man approaches unaware he is being watched. The officers hold back until he is completely out of sight of the event. The man, having hardly begun to believe his luck at avoiding arrest, was thinking to himself and reasoning that it was a charitable event and the season of charity and forgiveness, and even the cops are not so bad when you really get down to it, then suddenly as he turned the corner of the next block he was immediately dropped by 3 officers who put a gun right into his face and dared him to move a muscle.

In the meantime, Andy has moved into the spotlight position and is staring into the police camera, worrying that he had not bargained for this. He was always nervous around cops since being arrested as a young man on a few occasions for what he thought were trivial offenses. He was just there trying to get a little love from the nice but stupid man who was giving his money away, and half believing that he was going to be arrested at any time.

The police officer then asks him, "Are you Andy G. Clausen, of Warminster?"

Andy answers, "Yes, that is my old address!"

The assistant to Mr. Kennedy, standing next to the officer then asks, "Why are you here today mister Clausen?"

Andy now getting more nervous, starts to think to himself, "I need to tell him something that will impress them, that I am a needy person."

So he replies, "Well, I cant pay my rent and my mother is in the hospital, and my brother is in jail, so he cant help. And I would like to help my nephew finish college."

The assistant then asks "Why is your mother in the hospital?"

He replies, "She suffered a serious sickness?"

The assistant asks, "Can you elaborate on what her condition is? Is her life in danger?"

Andy then hesitates and says, "Well, yes she needs a heart replacement soon or she will die!"

The assistant then says to him, "What is wrong with her heart? Can you tell us her specific condition?"

Andy replies, "Heart, uhh, heart failure, and uhh, heart, uhh, grundage."

The assistant, laughing, looks at others with his group and discusses the described maladies with them.

He replies, "Did you say grundage? Umm, what is that? I'm sorry but none of the doctors present have ever heard of a condition called heart grundage. Mister Clausen, you wouldn't make something like that up to get money would you? You should be aware that we must verify all claims, which would include your claim about your mother's condition, before we could assist her, or you? I will tell you that if what you say is true, I can promise you that mister Kennedy will certainly consider your request for assistance. Do you have any other questions for me?"

Andy replies, "No sir! Oh, can you tell me when I will get the money?"

The assistant replies, "Well first you will need to make and application, and supply us with some details. Then we will investigate the situation and see if it warrants help from mister Kennedy. Now please step over to the table where they will help you with that."

Numerous additional events like this were conducted by Mr. Kennedy and his staff in the following years around the holidays, but not every year. They were reported on faithfully with periodic updates. Mr. Kennedy is subsequently dubbed, and is hereafter referred to as, 'The Philly Santa', and 'The Philly Saint.'

The succeeding years continued to see a larger turnout for this event culminating in the 5th event from then, with a crowd of nearly 5,000 requesters claiming status as needy, and applying for assistance from the Philly Santa. The event took 4 days to process, and he finally requested there be no more people applying, and had the police shut down the line.

Andy Clausen's claims about his mother needing a heart replacement were investigated and found to be without merit. He never heard from the Philly Santa, and never again attempted to extort money from him.

Neighbor From Hell

[MD Narrative] Rosa Cárdenas is occupying a house right next to my parents, the Senior Dalgleish's, whom were her previous employer in years past. She remembered being fired by them when they caught her stealing from them and mistreating their youngest children. I mean, really cruel and depraved things. My youngest sister was never the same after being tortured, beaten, and slapped regularly by her as a small child. They went to the police and the agency that sent her, and demanded an investigation into why she was given glowing praise in her references, only to find out the agency itself was under investigation for fraud, milling fake IDs' and references, and was being pursued by the state of Texas to shut it down.

But due to monumental bureaucratic bungling, legal red tape, corrupt government, and pressure from advocacy groups, this was near an impossibility and only fines were imposed with promises to clean up their act. Despite all this, and although she had only been employed by them for about 9 months, it was considerable trouble for her and she did not like it. In fact, the failure of both the regulatory authority to shut down the bogus agency, and law enforcement to prosecute or even deport her only emboldened them, and soon she was back employed by them working for other families for several more years. In her sick mind, she never forgot how my parents had wronged her by pursuing the matter instead of just ignoring it like she expected they would do. It may be, had they just left well enough alone, they would not have had the years of trouble they had from her, but then who could do that knowing what she had done?

He thought it possible she may have just done this in order to taunt them out of spite, or envy or boredom, because day after day she would hurl her venom at them from her perch next door. Day and night, when they or any of us came and went, she would be there yelling at us. I can only imagine what they had to put up with, especially since my dad could have turned all 3 of her CHARLIEs' against her and disposed of her castle at any time. I myself offered to issue a terminate command on her, and wish I had done so on many occasions.

§

Circa 2056, Michael and Boirix are discussing the problem of house stealing and property theft in general as they arrive in the front park garden lot of Boirix's house, in his autonomous car which is very large and ostentatious; a very common thing in Texas, "Does she ever shut up? Meghan is refusing to come over and bring the children anymore, because she believes your crazy neighbor is unstable. She is afraid she may be psychotic", says Michael.

There is no front or back orientation to the car, but it drives in either direction. Arriving in his neighborhood, they pass a few of his neighbors in their own cars, which are not CATs, but an assortment of various upscale luxury brands, still manufactured for the wealthy class, like Boirix's new Rolls Royce. Some have dual sets of steer horns mounted on either end, which is common in Texas.

They are in mid discussion about the irritating problem that has set itself up next door, as the assorted contingent of CHARLIEs' come over to greet them as they exit the car, when they encounter the familiar but irritating scream, "Hey meester dog-leash, or chewed I call joo dog-sheet, ees more like eet? How is misa dog-sheet today?"

Rosa screams and taunts from her second floor veranda on the side of the building. This perked the interest of Boirix's CHARLIEs' and several of them came and lined up behind where the two men were standing.

Boirix yells back, "Yeah, you too Rosa, why don't you just go inside and mind your business, before someone has to put a cork in that huge mouth of yours."

Michael pleads, "Dad, why do you take that crap from her, let me issue the directive, I'll make it dazzling, and spectacular, I mean, she won't know what happened or what hit her. How about I just scare the living shit out a her, have her CHARLIEs' come stay with us where she can see them, and put her on the prohibit list. Then maybe have them remove some or all of her house, I could do it in about 5 minutes."

"Why doan joo go e-site and geet out of my sight before I trow joo weet a rock. I coo smell joo. Joo stinks bad. Joo stinking up my air", Yells Rosa. "Joo think only joo why peeple coo leeve well, and no Mexicans. Well I cho joo this Mexican can leeve jo ah goo ah-joo."

Boirix says, "You know she's harmless, I think. She doesn't have the guts to do anything."

Michael asserts, "You are aware of the frequency of these kind of incidents to escalate, and she fits the profile to a tee? Aren't you concerned that she is being used by someone that wants to get to you, for Lord knows what reason?"

He replies, "Well you realize that we put a universal hedge directive around ours and yours, to prevent just such a contingency. I'm not sure how she got around it, or if Jim screwed it up and got the location wrong or what exactly, but notice that it did work to some degree because she was held back. The question though is still, why was she able to get this close? I'm looking into it."

Boirix continues, "Yeah, and you know it sure seems strange, she does not realize that her CHARLIEs' came from my company; you'd think she would at least suspect and realize how dangerous it is to start a war with me."

Michael replies, "You are assuming she has a rational mind; thats the difference between us; I would not make that assumption after what she did before. Someone is using her to get to you, can't you see that? Doesn't that bother you?"

He replies, "That is possible, but I don't see who, or what they might gain, and I cant live my life worrying about it, I have too many other things to worry about."

Michael asks, "Why aren't your security people doing something?"

Boirix replies, "I told you they are investigating, but I'm not worried."

Michael asks, "Would you mind if I looked into who is behind her move?"

Boirix replies, "Knock yourself out!"

§

[MD Narrative] I could plainly see that Rosa was driven by nothing more than old fashioned envy and jealousy, because she felt that we had been privileged all our lives, while her people were poor and lived as servants. Now believing that it is beneath her new station in life. Completely ignorant of the fact that we were very much responsible for creating the CHARLIEs' responsible for maintaining the prosperous new life she now had; simply dazzling ignorance. None of that mattered. Only that she felt she was wronged by her previous lot in life and deeply needed for someone to pay for that.

It was also so strange that she used CHARLIES. Where did she get the wherewithal to commission and keep CHARLIES? There were many of the upper class that only had one, where she had three, and she got them to ignore the protocol restrictions put on them, which were at the highest level, secret, and only a few people in the entire world had the access to modify them. They were universal and supposed to protect my parents and my immediate family domain from any damage or harm whatsoever, by all ARCH's and all CHARLIEs', but she got hers to build even on my parents lot but not destroy their house which she certainly would have done if she could have. This was a concern and a burning question I shared with my father, and I had to investigate. I knew she did not possess the intelligence or skill to circumvent the protocols; someone had designs on Boirix and was using her to play him.

I was more afraid for my parents at that time, and was very tempted to liquidate her myself, or just remove her from the scene before it was too late. I regret I passed up that chance.

Nuevo Luddito

[MD Narrative]: With massive and increasing unemployment, many were moved to revolt and take their revenge on the very machines that seemingly represent the sum-total of their problems. Massive rampages ensued and many machines were destroyed, or the companies that employed them were protested and boycotted. This however was only the inevitable violent reaction to the new way of doing things that always precedes the equally inevitable acquiescence. Since the time of the luddites who revolted against the industrial revolution and the proverbial obsolescence of the buggy whip, there has always been massive protest and anger followed by quiet acceptance when those on the short end of history finally realize their place in the new order.

§

The uproar created in the aftermath of the Belisle Whetstone affair did not start out small then grow, instead from the day after the event ended, there were numerous groups of very energized and loud people in large crowds that devised ways of mimicking or staging events of their own in the same vein as the Belisles' which they were dubbed by the media and several vocal individuals that gave statements about why it is they believed in Belisle, and what they intended to do in solidarity with the Belisles'. Some called themselves Belli-Warriors, and referred to the movement as the Belli-Wars.

The unemployed masses also see the corruption in Washington because of the robo-scandal, and how the political elitists are living large on what they claim is theirs; the graft which the politicians have always delivered to them in the past, and being unemployed they stage massive demonstrations for months, occupying large areas of the downtown and suburban sections of all major US cities; the same occurred in Europe and elsewhere. They went on rampages destroying the machines around them with elaborate rehearsed media productions in which they systematically dismember an entire group of CHARLIEs in what would be a gruesome and grotesque manner if it had been a human or an animal. Nonetheless people grimace, and cry because the machine's Fac-sentience feature has matured to some extent, which means they were particularly good at convincing the world of their faux-suffering at the hands of their torturers.

The population of the United States had grown to nearly 950 million by this time, and the unemployment rate had grown to nearly 40% of the working age population, some 160 million people in the past decade and a half. They staged very large events, with participants of nearly 3 million people in several major cities like Washington DC, Seattle, Los Angeles, Chicago, Milwaukee, Detroit, New York city, Atlanta, San Francisco, Dallas, Houston, Miami, Phoenix, Denver, and with very large participation in virtually all cities with populations of more than 5 million people; some 90 cities. These types of demonstrations also broke out spontaneously in virtually all the major cities around the globe that had a languishing industrial based population. It is estimated that nearly 1 in 6 people on the globe had participated in one form or another in these types of Belli-War events.

Even China which had never succeeded in supporting relatively full employment for more than 20% of their working age population, had massive demonstrations in which numerous representative samples of the vestiges of industrial automation were smashed, or destroyed in lively and colorful displays. China's political system had made some inroads earlier in the century toward industrialization, but was somewhat hampered within a few decades due to the wholesale displacement of employment in the developed world, caused by the uneven labor costs, and other imbalances that favored China at the disadvantage of their trading partners. This took its toll on trade with China, which eventually resulted in the slowing or outright reversal of trade with some, and created an uneasy artificial parity and stabilization as the trade door between China and much of the rest of the industrialized world began to shut.

The Chinese government was slow to let go of their death grip on the reigns of power and make necessary changes to maintain the trade relationships, but instead attempted to compensate with more automation and less workers. Even so, the Chinese people acquiesced to this as they had become accustomed to the social order arrangement defined by their rulers that could be characterized as a hive, complete with queen and worker bees. Regardless of the Chinese cultural temperament, the fact was there were many Chinese that were not happy with the situation and wanted change and were willing to brave arrest and the firing squad to make that point. It is estimated that nearly 300 million people in china followed the examples set forth by the rest of the globe and showed up loud and defiant, and took revenge on 10s' of thousands of examples of the machinery and complete factories, which they blamed as blocking the necessary progress for them and theirs.

In Chicago, there was immediately a particularly energetic group of Belli-Warriors led by an equally energetic personality, a brazen bomb-thrower by the name of Higuera Escobedo. They started by assembling in front of a large manufacturing plant employed in machining and manufacturing in the agriculture as well as the mineral exploration industries, called PrestonWard Fabrication & Machine. They assembled with picket signs and with megaphones stating their grievances to anyone that passed by or anyone that would listen, while handing out some literature that deified Belisle Whetstone and his brave crew for showing the way to get the attention of a national and even local government, deaf to their concerns. Not that it would do any good.

By most accounts, the situation got to the status-quo through many years of social and economic engineering by elitists who rigged the system, and that was not likely to change any time soon. Therefore these protests, although good for the spirits of those underemployed masses of humanity forced to live a life they viewed as beneath their stature, and who had seemingly been left out in the cold, they were considered as entertainment by those that view political events as a bloody spectator sport between contestants. They were reckoned as futile exercises by elitists; to quickly fade away and be forgotten. After all, there was a significant history of failed starts to these same kind of events that launched like a rocket, but soon fizzled; however, this time proved to be different.

Whether it was the proverbial straw that broke the camels back, or maybe the stars were aligned just right; or maybe it had to do with outrage over corrupt politicians and their use of machine butlers, the so called Robo-Scandal, or maybe the situation had simply risen to a critical mass of seething discontent. In all likelihood, it was a combination of events; several coming together at the same time; a confluence of event triggers. Whatever the reason, the energy was sufficient, so Belisle Whetstone was somewhat vindicated in sensing the correctness of the mood in his gamble, while uttering his immortal words, "The nation is with us this time. I really sense it."

Reality being as strange as it is, no one can believe that Belisle knew that his half-baked stunt would really amount to anything significant, but he was the match that lit the flame, that touched-off the fuse of an already seething powder keg of anger which was now just beginning the process of destroying everything that stood in the way.

Reporters show up in a location in Chicago, on the first day of the protest, which was the Monday just after the Belisle stunt to report on a fledgling protest. Some of them interview Higuera Escobedo and a few others gathered.

The female reporter, "Sir, can you tell us what it is you are trying to accomplish; are you connected with the events over the weekend in Milwaukee, and what is it you would like to tell the viewers watching?"

Escobedo responds, "Connected? No we are not connected to the people in Milwaukee, but we feel they have spoken what we also feel, so we are here to show that we are sympathetic with their message."

The reporter, "And what would that message be, can you be more specific?"

Escobedo replies, "I'll show you, watch! This is to make a loud and clear statement!"

The men are standing outside of the PrestonWard Fabrication & Machine company with 3 CHARLIE machines they picked up somewhere. Escobedo was let go from the company along with nearly 60% of the staff within the last few years. He was one of the leaders of the union of metal and fabrication workers. He directs some in the group to bring the CHARLIE units up and stand them next to the fence in front of the main entrance to the facility, directly in front of the company's sign. One of the group takes a high powered rifle out and shoots the CHARLIE mid-torso, which ruptures one or both of the 2 highly pressurized air-tanks, spectacularly destroying the unit with a sizable explosion. Another man takes out a chain-saw and rips into another one of the CHARLIEs', completely dismembering it. The 3rd unit was spared because the news reporter refused to watch or participate again. Afterward they aired the report of the event, showing footage of both acts of destruction which eventually made the national and global news. After the event, the organized group scattered back to their homes or hang-outs.

The next day Escobedo called the same reporter and asked her to come again promising something even bigger and more spectacular; she agreed, so the reporter crew showed up to find a large crowd of other news organizations and spectators already present.

The event starts when a few employees working at the company join-in with the organizers meeting the group outside the facility and bringing 14 confused looking CHARLIE units, commandeered from various parts of the massive facility outside on a flat-bed truck. The organizers along with the media are escorted into the facility grounds by the employee group. The organizers again, just as they had the day before, with great glee and showmanship, pull out a chain saw and slice a few CHARLIEs up for the numerous cameras pointed at the spectacle. This at once brings the management and the companies security streaming out into the staging area where they are met by a contingent of reporters and their crews, a large number of police, and a crowd of spectators anxiously anticipating a memorable event, all swarming the scene.

At this event, there are about 10 times as many people as there were the previous day who show up including organizers, sympathetic bystanders, police, the plant management, and the Cloud-cast news organizations. The police are alarmed and call for reinforcements in-case there is overt violence. Otherwise the officers let the organizers do what they are there to do, regardless that it takes place at a private facility and involves the destruction of private property. Four agitated members of the company management are present and loudly demanding something be done to stop the obviously planned action, including an extremely animated Trygve Kettilson. They complain that the organizers are disgruntled former employees bent on taking revenge on the company. The police hold the management back and allow the event to unfold. The balance of the CHARLIE units are directed to stand next to a classic mercedes automobile parked near the staging area which belongs to the well dressed Kettilson. He has come outside the plant and is franticly attempting to stop the event, barking orders and demands, at various groups and people.

He screams at the police officer in charge to do his job and arrest the lawbreakers, "I am the one that called you here. You already know they have destroyed company property and intend to destroy more. You are police, you need to stop this at once. They trespassed, broke in and took property, which they intend to destroy. Arrest them, take them all to jail, what are you waiting for?"

The officer responds, "Sir, don't tell me how to do my job, now back up, or would you like to go to jail? I don't know that they intend to destroy anything. We can't do anything unless there are laws broken, and I haven't seen any."

The officer comments to his colleague within earshot of Kettilson, "I'm kind of curious to see what happens; I heard these things explode if you shoot em right."

The belligerent and big-mouthed Trygve, then threatens the police accusing them of being duplicitous with the protesters, "You call yourself police? You are supposed to protect us. You haven't lifted a finger to stop this, which makes you part of them. If you won't do something, I will."

As Trygve is issuing his ultimatum to the local police, he does not notice that there is a large crane nearby with a CHARLIE at the helm, operating the large machine which has hoisted a large bulldozer overhead about 30 feet and positioned it directly over the area where his prized classic Mercedes is parked and surrounded by the balance of the amiable unsuspecting CHARLIEs'. One of the employees from the company is in communication with the CHARLIE operating the crane via a com-device, and although news cameras are directed toward the gaggle of CHARLIEs' near Kettilson's car, they are not necessarily looking up to notice what is going on. The man with the com-device walks over to where the reporters are gathered and directs them to point their cameras to the area above where the car and

CHARLIEs' are, then he gives the CHARLIE the signal to drop its load. The car and the CHARLIEs are destroyed at once; crushed completely; a spectacular site to behold.

Kettilson disappears then returns to the crowded scene a few minutes later with a large hand gun, which he fires into the air. The crowd ducks and scrambles for cover as he levels the gun toward the protestors, when he is immediately tackled and taken down by a few cops. He's hand-cuffed then led away into a patrol vehicle, screaming, and swearing in his broken english, vowing revenge.

The Cloud-caster news was sure to report on the event, so much grandstanding and many speeches were given for the media presence by both sides. The management of PrestonWard says, *"The plotters are disgruntled lunatics and criminals committing criminal acts"*, which is certainly true; they are mostly disaffected and disgruntled lunatics, while the lunatic protesters insist, *"There is a wider message and interest at work, which is bringing attention to the plight of a massive portion of the population dislocated from the traditional main-stream of society by the advent of labor-displacing automation. The political system has all but ignored the situation, so we must take it upon ourselves to do something about it."*

The media takes its traditional role of siding with the protestors and against the management of the company, making them look greedy and foolish, while sympathizing with the criminal rabble. They spotlight the fact that so many of the companies own current plant-worker employees are present outside and seem to have joined in solidarity with the protestors, risking their own jobs, which seem to be in very short supply, a point hard to argue against. For the most part, the public also sides with the protestors, viewing themselves as possible 'victims' of the same process. Many decide to join in the free-for-all, organizing events in their cities.

Starting the very next day there are several comprehensive reports of other similar events being staged in many cities in the US, and in Europe. Within a month, virtually all the large cities in the world are seeing similar events unless their law enforcement is able to put it down, which is very hard to do because of the shear numbers participating. Repressing these events just makes them grow larger as more protestors come out and join the others. The police and political establishment are visibly alarmed by their brazen and daring actions. There seems to be little restraint as machines are lifted wholesale from inside of companies or off the street, and ceremoniously, euphemistically 'slaughtered' in elaborately staged events. The corporations respond by locking down their inventory of CHARLIEs' and other equipment, then beefing up security, so the protesters simply resort to destroying other forms of private corporate property with similar flair. The media reports that a war, 'a belli-war', has begun and is increasing in size and strength day by day.

§

Belisle Whetstone is being held in a holding tank and is going to be arraigned in court on the Friday following his staged event. Although arrested by the FBI, he is not held nor charged by them, instead he is handed over to the local Milwaukee police, and courts. The feds stand down, fearing a larger problem may erupt if he is deep-sixed. He is taken into court while hog-tied and thrown face down in a prone position, bent over a mechanical platform called the 'Iron Princess', which encages his body with his head and feet down near the floor and his rear-end sticking up. He is before a vicious female judge, The honorable Judge Willetta Schmauser, affectionately referred to as the 'Iron Princess.'

She lectures and scowls at him, "I caught your little stunt on Milwaukee Cloud-View Television, and I can tell you, I was not amused. You had better not trifle with my court, because you are already in a mountain of trouble. Seems you may have touched off a global riot; starting it in my city."

Belisle looks up and notices his covered mound poking through the bars of the iron-princess and situated in a rather vulnerable position. The judge brings up the charge sheet, and while looking it over on the tablet, she mumbles to herself then reads the list of charges, "Kidnapping, grand theft, terrorism, public endangerment, starting a riot. Quite a list of charges, and that is just the beginning."

With Belisle's attorney, Mr. Bigelow, standing right in front of her, she crassly says, "I hope you can find a good lawyer son, because you are going to need one!"

She asks Belisle's counsel how he pleads on each and every one of the charges which are 7 in all, to which his attorney, Rufus Bigelow, Esquire, sweating profusely, replies, "Not guilty your honor. The charges are not valid, but trumped up by the police. I would ask the court that the charges be dropped and my client set free."

The judge then orders the bailiff, "Bailiff, bring up another seat of honor and install mister Whetstone's counsel, mister Bigelow, esquire, in it."

The bailiff complies and puts Belisle's counsel in it as if he is sitting upright in a very prone position about 2 feet below the scowling judge with his butt in a hole, and his legs and torso sticking out.

She again asks Bigelow, "Lets try this again, how does mister Whetstone plead?"

Judge Schmauser was obviously referred to as the 'Iron Princess' because of her use of these mechanical apparatus, and the fact that she affectionately referred to them by that name. The devices, when in operation, make the honoree sit in such a prone position so as to accommodate the male anatomy in a rather uncomfortable way, giving the distinct feeling of vulnerability as the array of situated levers, clanking metal gears and such, moving about, seem as if they could easily compress or dismember parts. That along with the castrating and harsh demanding tone of the judge, gives the honoree the distinct fear that something unpleasant could happen upon a wrong answer, or the slightest hint of flippancy, and so is very persuasive and effective in focusing the mind, gaining truthfulness and the fullness of attention. As with Belisle, counselor Bigelow is keenly aware of the situation vis-à-vis his member, and the precarious position he seems to find himself, as the array of whirling and grinding parts seem to operate mere millimeters from his business, giving him a distinct motivation for truthfulness and the utmostedness of respectfulness.

A nervous and sweating Mr. Bigelow-Esquire gives the judge the exact same answer again, "My client is innocent of all charges your honor, ma'am!"

She then orders the bailiff, "bailiff, I want you to hoist mister Whetstone's attention my way."

The bailiff operates petals on the device that flips Belisle over and lands him in another prone position, butt first in the hole of the Iron-princess, with his head upright, hands still bound together.

She then asks Belisle, "Why on earth would you answer with such a croc as that? Innocent? We shall see!"

Belisle replies, "Your honor, if you will, please look at the footage?"

The Autonomous Economy I - The Strong Man

The lawyer hands the judge his com devise with some of the footage left unreported by the media. It depicts the 2 ostensible hostages as they are briefed on their roles by Belisle just before the stunt.

This relaxes the judge a bit, who comments, "This puts the whole thing in a different light. I am inclined to accept your plea. Mister Bigelow, I believe your plea but your client will still have to answer some charges."

Belisle replies, "I accept that your honor."

The judge and counselor Bigelow begin to discuss the charges. She agrees that most of the heavy charges are bogus and trumped, and so he negotiates the 3 remaining down to misdemeanors. Eventually she is left without anything to hold him on except petty stuff like disturbing the peace, destruction of public property, and creating a public menace, which he pleads no-contest to. Even a charge of grand theft is reduced based on a lack of evidence or complaint, so she lets him and the rest of Belisle's misguided followers go, but advises them, "Don't come back into my court again mister Whetstone or next time I may not be so understanding. Pay the fine on the way out."

Belisle walks down the courthouse steps and is greeted by a massive mob of reporters and well wishers chanting his name, so he stops to say a few words of encouragement to his throng, "This movement has barely gotten underway. We have a big fight ahead, and I am counting on you to keep it going and to bring it to where it hurts those that are hurting you, and remember that I will be there with you. Thank you for your confidence, I am truly honored by you."

Then his lawyer, Mr. Bigelow wades hip-deep into the throng of media, in an attempt to get as much free publicity as he could, "... thats Bigelow, B, I, G, E, L, O, W, don't forget, Esquire...", on and on.

The feds are interested in charging Belisle, but see him as a key to a situation that is really starting to get out of hand with regard to the demonstrations he initiated earlier, so they picked him up and brought him in to FBI Agent Brian Girardano's office to question him. They threaten to charge him with a myriad of charges, and the political reality of the day being what it was, they were certainly capable of finding a reason for successfully deep-sixing him. They offer him a deal if he will work for them, which he refuses, but relents with a tenuous offer of assistance motivated by his own desire to see his big 'master-plan' yield something other than the anarchy, social upheaval, and destruction it has so far.

Agent Girardano, "You unleashed something, of which, I am not sure you are appreciative of the magnitude. We have enough on you for your stunt to make sure you never see so much as a sliver of light ever again, and, we can go back to some of your previous stupidity. We hardly need to drop but into the shallow end of the lake to dredge up some bodies with your fingerprints on em. We can charge you plenty. You have to appreciate that mister Whetstone, and I am not sure you do."

Belisle replies, "If you had something, you would have already charged me. You don't have anything, or at least charging me is not what you want. You need me, don't you? I threw a match on a mountain of smoldering debris which erupted, and threatens to burn down the world. You need me to help you put it out."

Girardano replies, "Don't flatter yourself; you're not as important as you believe. The only reason you are not in a federal facility right now is that it may serve a bigger purpose with you on the outside for the time being, but it really makes no difference to me. I'd just as soon toss you out with the rest of the garbage. Let me warn you, don't make any plans to go anywhere outside of the area, and you had better be available if we come calling, or we might just dredge up some of those bodies."

They let Belisle go, so he gets back with his group and they formulate messages in an attempt to guide the movement, which are received very readily by the raucous hordes of rabble.

Presidenté Fullerty

Within a month of the start of the belli-war demonstrations, press reports and commentary start to proclaim general anarchy. President Fullerty gives a press conference address to speak about a general state of unquiet and civil disobedience, while calls for marshall law and troops in the streets are threatened if the lawlessness is not brought under control, which is laughed at and only emboldens the insurgency. The government looks very weak and feeble, and not in control of anything. The government had always acted as enforcer for the minioned-class, implementing and enforcing the dictates of the cabal, and this situation somehow did not seem to fit the usual expected pattern.

The usefulness of virulent riots and mayhem notwithstanding, non-sanctioned violent movement without direction or purpose is counterproductive to them. It spread like a wave across the US and threatened to consume everything in its wake. It took on the dimensions of low-level anarchy because it was not the usual contrived 'incident' so useful to the brotherhood. This was real and explosive; genuine anger freely released; the dam bursting kind. The kind of thing that in earlier days, the brotherhood had occasionally unleashed. The kind which led up to the French Revolution, with the rolling of many heads. The anger flowed alongside rivers of blood until it was finally quenched, having consumed those that engineered the revolution. Since those days, they had learned a more judicious and measured use of power, but the current situation was not expected, unleashed, nor sanctioned by the minioned; neither was it understood by them. It was spontaneous, genuine and 'grassroots'; a massive release of anger and frustration that had built up for decades. The psychology of the angry crowd is somewhat mindless, like in war; like a fight for survival; all that matters is getting what they believe is due them. Individuals act in ways they ordinarily would not under normal circumstance in which the power of restraint and deterrents are active and present. Individually, they took the lack of a reaction or crack-down as license to escalate, and used the crowd as cover for their actions.

Presidenté elect Fullerty was now Presidenté Fullerty and although many, including the minioned-brotherhood looked to him to deal with the fury, he was less than adequately initiated to the post to be of much help in stemming the wave of destruction. Whether it was his lack of experience in wielding power against his countrymen, or maybe he had not completely reckoned his status as marionette-in-chief, so was suffering moral doubt, acting in compliance with the cabal; in any case, he was not up to the task. In past incidences of this magnitude, the government was quick to crack-down on the rabble, instead his hesitation and feeble response only reinforced a quick and clear go signal.

Fullerty steps up to the microphone to address the situation,

"Thank you all for coming. Let me tell you a funny story. As you all know, I have barely gotten into town and found my way around. My wife says she likes it here so far, and the kids are excited. You'll be glad to know that we've managed to find out where the kitchen is and a few other necessities, and we were just starting to get settled in when..."

As he is speaking, he is interrupted by several press staff asking questions all at once, which is confusing and daunting for him; aggregating the storm of questions to, "Mister president, can you address the growing tension; the terrible rioting and violence. Some are calling it a state of anarchy. The nation is waiting. What is your plan to stop it?"

He continues, "I haven't finished my story. Don't you all want to hear my funny story?" With a goofy and very undignified grin on his face, he hesitates and peers out over a very stern and somber looking crowd that is not smiling back.

He him-haws a bit, then with a cracking low voice he squeaks out, "I have been in consultation with my staff..., umm..., some of them seem to think we need to study the options..., uh..., emergency power..., and..., measures..., including marshall law was suggested by some of them, but that seems a bit extreme. If you were to ask me personally..., you know..., umm..., what would I do to address the violence..., well..., I think...., personally..., umm..., that umm...., we need to talk to them, to find out what is really bothering them. Then...., umm..., I think we can lay a basis for addressing the umm..., underlying causes of..., of, of...., anger!. Next question!"

The next flurry of questions could be aggregated as, "Mister president, what do you have to say about the robo-gate scandal, will you call for an investigation?"

He walks away dismissing the press conference after hearing Senator Schuman sitting off to the side audibly say, "Will someone tell that green idiot to shut-up and sit down."

Hegemic Polemic

[MD Narrative] The minioned-class's optimism about the future stems from the belief that as automation technology matures and the machines gain ground, and increasingly perform the knowledge-base and skill-base of the human population,

their dependance on the vastness of the human population to sustain them will decrease. While so, they have a deep seated phobia, and feel threatened as the population of the earth continues to grow at accelerated pace. The same era which promises to unshackle them from their dependence, they fear also threatens them, because the human population also becomes less dependent on those things which the hegemonists monopolize and employ to control all the affairs of the planet. This is a dilemma, and a threat to their control, and their anticipated exclusive utopian landscape.

§

Addressing the panic set in with less than satisfactory answers to concerns befalling the Hegemonists during the current session, when members began to display skepticism about the outcome of the era about to unfold, both Belshazzar and Xerxes are alarmed at the mood and tenor of the questions, so resort to speaking in terms of agreement and not adverse, in order to address the panic surfacing.

A man in the crowd stands and proclaims his fear that things will get out of control, "The situation is unraveling. This is not what we expected. They should've accepted the situation as it came, and not revolted like a pack of wild animals. There is only one course, which is to appease them, and that puts a boot on our necks!"

BelShazzar retorts, "We are the authors of unrest. We spur riots and turmoil everyday; why are you worried about this situation. There's no need for alarm. I don't share your angst, nor do the others here!"

Another one asks, "Again, If money, and profit disappear, how do we maintain control over anything? Why should we trust what you say. You have not answered the concerns. Again I say, this situation is not good for any of us, until you can show otherwise!"

Xerxes retorts, "Please, please, those of you so concerned, lets not panic here. Its not in keeping with people of your stature and position, nor this organization. We are all still alive, and I can assure all of you, we are in control of the situation. We do not know this Belisle, and doubt his ascension to our ranks. We believe him to be a usurper. He is a flea; do not concern yourselves over this matter."

BelShazzar speaks and addresses the concern with rhetoric designed to restore confidence by using language that appeases the bloodlust of those present, "With regard to Belisle, it is characteristic of my opponent, that he should err in this. Of course he does our work. Indeed he is insignificant, he merely puts the incident of our work to effect. Now, lets examine the facts shall we. Fact, The vastness of goy have served our purposes from the beginning, even till now! Fact, we appear to be entering into an era as foretold, and I, no we, without question believe it to be so. This era is for us; they are not a part of it. There is only one way to interpret that, which is, they will eliminate themselves, and we will remain, however, our assistance may need help the process."

He continues, "Fact, in the past we have always confounded and exploited them by several methods. Let us examine our own protocols. We have always fostered problem and crisis; for us, solution. Through cultivating the baser elements of their own nature; through political theater. We sift them like sand and order them as it suits us. We direct, so, we have used their institutions, their civil and political structures, their economic systems, their courts, their media channels, their religion, their universities and primary schools. We have compromised their political systems to a mere game of democratic pretense. Their governments have been completely co-opted by us. They have allowed all of their prized and public institutions to become a haven and abode for those not burdened with honesty nor conscience, who readily enrich themselves through the rivers of graft, according to the axiom, '*A box of shit attracts a lot of flies.*'"

He continues, "Let me read from the chronicle. We have commanded the entirety of their social structures. We have so fractured the landscape of the identity of civilization."

He interjects, "mostly in the advanced western world; with global infusion of multicultural cesspools from competing and warring cultures."

Continuing again from the chronicle, "What were once as a strong and clear pane of pure glass, is now webbed and laden with such immense internal stress, so that it takes a mere tap from our small mallet to shatter the whole, never to be put together again. There is no single system; no point of power or cohesion with which they may muster the power of challenge. We have woven such a fabric of diverse and disparate interests who compete and war against each other; they have no resolve. We have so destroyed the once homogenous foundation field of pure nation, now sewn with the

325

indelibly intrenched predatory weeds and bramble of all others, rendering the nations impotent. Their focus is diverted from us. Their energy is wasted with infighting so as to resurrect and maintain their long dead dream of ethnic national identity. They can no more resist us, than arrive at a consensus apart from that which we ordain."

He puts down the chronicle and continues, "In the same fashion, we have ordained in their systems of law; burdened by welfare and entitlement; for one and many abjectly non-productive racial groups having been made dependents of the whole, and the whole being legally obligated to the burden, so as to create a climate of racial resentment, while heaping un-payable debt upon a nation, until the internal stress reaches critical mass. Their parasites do not hold back but continuously make shrill demand of the others, and many fools among them defend the folly, believing themselves virtuous in their magnanimity with public money."

"With murderous intent, they preach tolerance; indeed, they have elevated tolerance as the highest societal virtue, against any deemed intolerant; that, non-repentant, should suffer and pay the ultimate penalty; A delicious reigning ethos of contradiction. We have implanted in the collective mind that right is wrong, is left, is right; up is down; over is under is inside and out; good is evil, and evil is good; therefore 2 and 2 equals 9; now the only things deemed good and embraced, are those believed irredeemably evil only a generation before, and the only things condemned, are those held holy and sacred a generation before. They accuse and indict their fore-bearers; the very pool from which they spring; their very origins and essence, as corrupt and reprobate. Now they do our work."

"The best part is, we have done these things while making them believe that these are their own doings. They believe they have created what in reality is an unstable house of cards by their own craftiness; refusing to acknowledge the terminal flaws. They have forgotten the past, when civilizations were strong and based on governing men in harmony, within the sound dictates of nature, which preserves and affirms life. A time when an empire was a rock of certitude in a sea of turmoil, and could last a thousand years."

Continuing, "Fact, They believe they have arrived at the apex of civilization; the optimum achievable. They stare directly at us, as we murder them, and see only a shadow. They believe they alone have engineered their backward civilization based on optimum principles. They believe their undoing is their own doing, and refuse to recognize the spoiler in the midst, who has compromised it at every turn, to the extent that it is now, unfit, and only worthy of destruction. Should an angel ascend to earth's surface and demonstrate to them the spoiler has unquestionably destroyed their civilization, they would not believe such is possible."

"We have cultivated vile appetites among their masses, weakening them; fostering their regression to that of animals. We have magnified the finger of accusations on all sides as coming from the others, exploiting divisions, fostering war. This has always given us the upper hand. When crisis has inevitably risen to the frothy peek, we have delivered the means for all sides to solve it in the manner they invariably choose; animal killing animal."

"All-things, in their time, has always been an important factor for us. Without our restraining them, they would've already destroyed themselves. Its a simple matter of scale, and giving them the correct tools at the proper time. By these things, and many, many more, we have always garnered the power to do whatever we may, whenever it suits our purposes. Without exception, this has worked for us and this time is no different. We know what we do. We have learned many things; we have a very good mentor who's been at it from the beginning."

The same agitated man, still standing, interrupts, "All of that does not matter; their weakness will not matter, but their instinct for vengeance will matter, and remain intact. It does not matter if they know who we are, or if they guard against what we can do to them now; they will remember what we have done and they will come after us. The time coming, transcends the futility and ignorance of the human race. Your point is that we have always exploited their ignorance and self destructive tendencies; that only works if we control what they need to survive. Again, it won't matter then; we will not hold sway any longer. If we cannot control them they do not need us, and if they do not need us, we cannot control them. We must stop this from happening; we must stop this age from descending; this is not our time, this is our end."

The chairman of the panel, Sargon, out of order, interjects, "We have heard enough! Let us take this under advisement. We are adjourned for 3 hours and will return with our counsel."

Upon the counsel's return, Sargon announces, "This is the first time, at least since I have sat in this chair, and it may be unprecedented. We have not reached agreement according to protocol, therefore, in the absence of consensus, protocol dictates we table the issue to a later time. In the interim, we will designate resources to further study the issue and devise measures adequate to eliminate any ill circumstances. So say all! So said, so agreed, so sealed, so done!"

Despite the reassurance of the leadership, many doubt the power of the worlds elitists to maintain control in the face of what is coming, but many more kept their faith. Believing BelShazzar and others do not see the writing on the wall which is quite evident to them, they secretly settle their business, cash in their chips and make plans for the survival of what they believe to be a disaster about to befall the minioned-class worldwide. Their actions are based on moral stories; that one will reap what one has sewn, and their particular crop is immense. Seeking to preserve themselves and their families, they take heed and go into hiding, fearing the fallout if the worst case scenario comes to pass. They are not seen again amongst the minioned-class gatherings, which in itself is considerably unsettling to many of their comrades.

In succeeding meetings and plenary sessions, there are noticeably fewer participants of even those considered stalwarts and pillars. Now deliberation is replaced with considerable squabbling and infighting, while the so called belli-wars continue to rage and gain strength around the globe.

§

[MD Narrative] The hegemonists knew the gamble of things getting out of hand if the population were to realize their own status in an autonomous economy, and they were aware that such a state may be approaching. They feared the 'rabble' would discover they can survive without the 'grace' of their overlords, thus the dilemma. This figured large in the dialectic discussion.

Significant numbers of their ranks suffered from a particular psychosis, engendered by unearned wealth and power, extreme corruption, and unaccountability, and it drove them with an emotional appetite to dominate other members of the human family, so they had planned for a limited depopulation of the earth in order to preserve a certain number of goy-minion that can be humiliated and dominated by them personally.

§

It is clear they feared the contingency, and were vulnerable despite their belief of not showing vulnerability otherwise. Some in their ranks had an epiphany and discovered they were, in reality, not all-powerful, nor invincible, and their eventual, inevitable, complete, uncontested dominance, was in-doubt. They were apprehensive, as well as hopeful of the future and what it may bring as they approached a crossroads, which only proves them to be as human as any other. On the one hand, they may see the advent of what they have worked and planned for millennia, and on the other hand, they could very well loose all.

Pearls and Swine

On the last such update, the man called the Philly Santa is seated in his usual location, in his gilded, bejeweled throne, and attired with his usual flair. The usual crowd of media, hopeful needy recipients and those looking to cash-in on the rich fool is bigger than anytime previous. As usual, they are lined-up around the corner for the event.

This time though, he is alone and has no one assisting him so he addresses the crowd directly, "I am sorry to have to tell you this, but my fortune, which at one time hit 700 million dollars is gone. It had been substantially derived from an investment into Dalbots and other equities, but, due to global economic de-leveraging, the loss of business and the economic disruption from global rioting, along with a massive tax increase on corporations and wealthy individuals and generally declining stock prices, my debts have exceeded my fortune. I am now broke. This means I can no longer provide assistance, any assistance, to anyone. Thank you for understanding, Good luck, and God Bless!"

Upon hearing the news, they are aghast and shocked. There is heard the low dull roar of rancorous murmur which rises to a crescendo with wailing and angry shouts. Many screaming and whaling out-loud with writhing and grimacing faces and displays of absolute poetic drama as they show their disdain for the previously hailed, Philly-Saint, now derisively called the Philly-Flimflam, the Philly-Con, with unflattering epithets and insults. As the crowd starts to disperse, the ever-present ruthless gamers and hustlers, with taunts and screams, begin demanding the man be arrested and prosecuted for his unforgivable crime; not being careful enough to watch and preserve his fortune, so he could manage to care for those he was now obligated with for care and salvation. Many others in the crowd also chimed in.

Even the news reporter and the anchors let loose their disdain, publicly shaming the man in solidarity with the screaming ungrateful mob, threatening the man with bodily harm, "This is horrible. We are witnessing the abandonment of these needy and desperate people by a man not responsible enough to preserve what so many have come to rely on. Absolutely shameful. I am being told that the state attorney may look into bringing charges against mister Kennedy, for dereliction of social responsibility. The 2037 world convention on human rights has called for it, and global law requires it. He has obviously been frivolous with what has been entrusted to him. Now the innocent are suffering for his irresponsible actions."

They watch and report, commenting on his beating, as a crowd of 8 to 10 individuals take ahold of Mr. Kennedy. One of them sucker punches him, knocking him to the ground, breaking his teeth. Two others grab his coat, ripping it off of him, then kick him repeatedly in the mid section, then go through his pockets taking whatever he has, while he lays motionless on the ground. Meanwhile the crowd is fighting over his beloved 'throne' chair which is elaborately carved and decorated with inlaid precious stones, gilded with gold leaf and fine detail and upholstery. Several of them pull on the chair, attempting to take it from the others, breaking it into pieces. Then each of the vicious mob takes a piece and runs off with it in different directions.

Texican Standoff

[MD Narrative] I distinctly remember these events. I have never in my life been as terrified as I was during that time. There was tremendous fallout as massive riots, protests and demonstrations ensued. It threatened my family, my parents and everyone that mattered to me at the time.

§

In Dallas Texas, there was a growing presence of the belli-war movement. They had decided to focus their attention on what they identified as the very source of their problems, in attempt to storm the Dalbots headquarters itself. Boirix called on the Governor of Texas asking for protection. He also calls General Cranket for assistance.

Texas, which had always maintained a kind of defiance, in your face, frontier, outer fringe, outlaw, character of tolerance for individual eccentricity and quirkiness, and the baggage that those which brung it trod behind them; but tolerate not, nor suffer did they, stupidity, nor abject vulgarity, nor violence, nor open hostility toward them and theirs. You were welcomed into the big ol' Texas family as you come, but with an attitude of, *'mind your own biness, and keep your nose clean, or we'll be settin you out at the curb!'* Most of Texas was Texas-normal, which is much like America-normal, but with a Texas flair, so most of the more peculiar element were concentrated, and some would prefer 'penned-up', in small oddball enclaves. Many were homespun, but many more, maybe most of this crowd came from areas outside of Texas or even the southern United States, from far and wide, to set up state residency and blend into the backdrop, forming an assortment of colorful rabble.

When the belli-wars finally focused their attention on Texas, Dalbots, and Boirix, many of the local rabble from the area joined in with the large mass coming in to protest and create havoc, to the extent that between the resident fringe rabble, and the many, many, that came into the area from outside, the effect was massive, with protests and rallies in the streets of Dallas and other cities, disrupting, blocking streets and biness, destroyin property, or just interfering in the biness of the locals, angering many a Texan that was not all that happy with the state of things, or the increasingly venomous calliope of blather from the rabble.

The outsiders were seeking attention for their cause on Texas soil, by trampling on the good name and reputation of what many a Texan saw as representative of home pride and one of their favorite sons. Action was needed, and a good Texan does not need to be asked more than once when duty calls. The locals rose up to the challenge and showed the homespun as well as the imported rabble a few manners, with greetings like the 'Happy Texas howdy-do', and the 'Happy Texas farewell', both of which are usually given one right after the other, and involve five dudes with a set of persuasion tools, like a number 5 Louisville Slugger, in full dress with hat, buckle, with size 11 or larger high grade leather pointed shit-kickers, with reinforced steel tips, skillfully placed right up old brown blinky, some in Texas call the 'under-eye.' These tactics and more resulted in considerable numbers of the ranks of recently arriving rabble decidedly persuaded to move-on to more fruitful venues for their rants, however, it was not enough to dislodge them altogether, while the majority remained, and more kept coming.

The streets are blocked-off by the locals around Dalbots, a significant and obvious target and focus of the rage spilling everywhere. There is palpable tension in the air for the better part of a week, as the throng move enmasse toward the area. They come in CATs, aircraft, or just move on foot, leaving a scorched-earth in their wake. Like a plague of consuming locust, they take whatever they need and occupy wherever they find themselves, stripping store shelves bare, appropriating whatever they need regardless of who objects or attempts to protect their property. If avenues are blocked off, they remove the barrier or simply go over or around it. Estimates of the throng descending on Dallas were in the 1 to 2 million range.

A historic bloody confrontation was shaping up, the dimensions of which promised to reshape civilization. A solution is needed to return the peace and end the stand-off.

Appendix

Terms and Definitions

Entropy
The term was derived and applied to the field of physics by the German physicist Rudolf Clausius (1822 – 1888) and based on earlier theoretical work by others. The word 'entropy' comes from the greek, *en* (in) and *tropē* (transition, transform), which refers to the amount of energy lost or gained in the transition of a system when work is done. It is the measure of the energy contained in a system that can be used to perform work, and related to the efficiency (loss of energy) when work is done. It also refers to the measure of the degree of order or disorder in the universe. In a thermodynamic system all processes tend toward the loss of energy which is not recoverable without expending more work. Energy is simply lost to the environment.

It is not a perfect analogy, but close, when used to describe the fall of mankind in the Garden of Eden, but it can be abstractly applied in the context to describe the state of an economic system, which can be expressed in terms of energy and order. In the Garden of Eden,

The Autonomous Economy I - The Strong Man

Adam & Eve lived in a perfect world, so by application, there was no entropy; consequently there was no cost to them for sustaining their lives, wheres in the fallen state in-which humanity currently lives, there is always a profound cost for the same. This idea can also be expressed as achieving *unity* or *over-unity* energy generation, and other utopian ideas about the economics of sustaining life.

In the garden, a price was not required of Adam & Eve to sustain their lives, until after their disobedience. The Garden of Eden was a paradise in-which there was no cost for them to eat, live, and enjoy their lives, and that state would have been maintained forever.

After the fall, when entropy reigned in their environment, there was always a loss in terms of anything they did, consequently, they were put at a tremendous disadvantage and required to pay a profound price for any advantage employed in overcoming the harsh demands of nature. God pronounced that Adam would pay the price expending work (*in the sweat of thy brow*), tilling the ground to bring forth bread for them to live, and Eve would suffer pain in childbirth, and because they were now under the curse of death, and would eventually die, children were necessary to continue the race.

Ultimately they lost the battle when the elements overcame them and they died, thus fulfilling the warning of God who told them that they would die if they ate of the fruit of the forbidden tree. It is a simple suggestion that after their fall, the memories of their perfect lives in the garden was the *Knowledge of Good*, whereas after the fall, learning the skill for survival and becoming familiar with the price required for overcoming the elements to sustain life, can be thought of as the *Knowledge of Evil*.

Boirix

Boirix or Boiorix pronounced phonetically Bweer-icks - Bwaur-icks - or Bwyor-icks. King of the Boii, from Rix (king), and Boii (an ancient 2nd century BC through the 1st century AD, Celtic tribe that dominated Northern - Central Europe (Bohemia), and coincidentally, also dominated the Cis-Alpine - Northern Etruscan peninsula (Bologna Italy). Their tribal name (Boii) was thought to have come from the Gaulish term for "Cattle Owner", a designation of Aristocrat, as the ownership of livestock was evidence of wealth and power for migratory warrior tribes.   Boiorix, although somewhat obscure, the most notable figure to bear it was the leader of the Cimbri tribe of Jutland (Denmark).   In this case, the name was borrowed, or adapted by the hybrid Celto-Germanic tribe, after the original tribal namesake had fallen in stature somewhat.

Dalbots

Dalbots is headquartered in Dallas Texas. Founded in 2019 by Boirix Dalgleish, with 2 co-founders, and an investment group. Part of the initial startup financing came from a US government grant. The name, 'Dalbots' is a mix of Dalgleish and Robots, and not a contraction of Dallas and Robots, as some mistakenly believe. Dalbots makes automation equipment, used in retail and industrial settings.   The main product lines include the ARCH6s' and CHARLIE series, autonomous Humanoid Robots; Robotic kiosks (obsolete); self-help/self-serve station systems for the retail industry (discontinued); and a limited military division, producing a limited line of classified OEM robotics equipment.

The latest technology (starting circa 2030) from Dalbots is the ARCH5, ARCH6, and CHARLIE series systems, which are autonomous humanoid robots.   Equipped with robust advanced control systems, they are artificially-intelligent and adaptable humanoid robots, generally tasked as general purpose machines, suitable for domestic and industrial labor, and work in hazardous environments not suitable for humans.

ARCH

Short for ARCHIE - Models are the 540, the first full prototype in the limited 500 demonstration series, and the 600 production series. All are advanced humanoid robotic models.

Dalbots ARCHIE for ***ARtificial Computerized Humanoid Intelligent Entity***. These are intelligent 'fac-sentient™' machines that exhibit advanced artificial intelligence. They are able to converse, interact, and generally behave well among humans.   Humans have come to trust and rely on these machines, and employ them extensively. Early models of these machines start to appear in the mid 30's.

CHARLIE

CHARLIE - These are 2nd generation Advanced Artificially intelligent Humanoid Machines.

Dalbots CHARLIE for _Computerized Humanoid ARtificialLy Intelligent Entities_. Like the ARCHIEs, but enhanced, and more advanced version that starts to appear in the mid to late 40's.

Fac-Sentient™

Fac-Sentient™, or Faxsentient™. Apparent but artificial, software-simulated sentience. This feature makes the ARCH6 and CHARLIE seem alive. It gives the machines a more human quality, though not in the real sense of self-awareness, nor introspection, thus enhancing the human trust factor. Sentience is the aspect in humans and some animals observed as feelings or an emotional seat. Many jokingly refer to it by the name 'Faux-sentient.'

Singularity

The word **singularity** was derived in the fields of physics and mathematics to describe an observable but unusual phenomena, which may not exist in the real world, such as a single dimensionless point or something in the infinite, meaning not physically bounded. A dimensionless point would not usually be encountered by the casual observer in the real world. In physics, it is used to describe the advent of an 'event horizon' at the formation of a black hole.

The word has been adopted by fiction writers and futurists, such as Ray Kurzweil, to describe some catalyzing event which plays a significantly transformative role in human history, and tending toward extremely rapid and unlimited change, such as super intelligence in computers. In _The Autonomous Economy_, the word is used to describe an event that is significantly different in effect and outcome than the meaning by other writers and futurists. In the Autonomous Economy, the author describes a rapidly changing transformative point in history (future history, as told by a character in the book) in which technological advancement in the area of machine automation is replacing human labor, and its effect on the global economy, which has transformed the way in which humans live and interact.

From: http://en.wikipedia.org/wiki/Technological_singularity

Technological singularity refers to the hypothetical future emergence of greater-than-human intelligence through technological means. Since the capabilities of such intelligence would be difficult for an unaided human mind to comprehend, the occurrence of a technological singularity is seen as an intellectual event horizon, beyond which events cannot be predicted or understood. Proponents of the singularity typically state that an "intelligence explosion" is a key factor of the Singularity where superintelligences design successive generations of increasingly powerful minds.

This hypothesized process of intelligent self-modification might occur very quickly, and might not stop until the agent's cognitive abilities greatly surpass that of any human. The term "intelligence explosion" is therefore sometimes used to refer to this scenario.

The term was coined by science fiction writer Vernor Vinge, who argues that artificial intelligence, human biological enhancement or brain-computer interfaces could be possible causes of the singularity. The concept is popularized by futurists like Ray Kurzweil and it is expected by proponents to occur sometime in the 21st century, although estimates do vary.

Note the author, CP McCollum, was unaware that the term had been adopted or applied in the context described by other authors and futurists, when he independently adopted the same term to name the first chapter of 'The Autonomous Economy', referring to the concept that describes a similar phenomena of unknown events ranging past a critical trigger point.

X-Pack™ Facility

Generically Known as the X-Packs™, software packages like Kno-Pack™, Do-Pack™; they also include Persona-Pack™, Work-Pack™, Proto-Pack™, and others. They are trademarked expert knowledge packages, expert skill, persona modules, work applications, and blank templates optimized for adoption of expert knowledge. Dalbots, and other 3rd party software modules are contained in a supervisory execution and security wrapper, specially developed and patented by Dalbots. They can be installed and run in ARCHIE5/6 series and

CHARLIE units, giving them expert knowledge and skill to perform specialized tasks. The Persona-Pack™ allows for 3rd party, custom configuration of personality, apart from the factory standard. Work-Pack™ facilitates general applications, and development execution.

Proto-Pack™ versions are optimized blank templates that facilitate the cumulative acquisition and compilation of data and knowledge programming for building commercial, and general purpose X-Packs™. X-Packs™ succeeded the earlier Dalbots SoftSnow™ standard and packages, built on the older pre-Dalbots generation of smart kiosks, which were phased out. There are numerous competing packs marketed, but all Dalbots machines only accept installation via an X-Pack™ wrapper, so must comply with the Dalbots open API (Application Programming Interface) standard.

QCD and QICD

Quantum Computing Devices - QICDs, and Quantum Influenced Computing Devices - QICDs. Computing devices based on the principles exhibited in the Physics of Quantum mechanics in which computational elements can be harnessed using superposition or 1,0, and 1<=>0 states. Quantum based computers are integrated and operate on superconducting material substrates, thus requiring exotic cooling systems involving liquid hydrogen or other coolants while QICDs require less exotic systems for operation, and are less vulnerable to noise and magnetic interference, but are slower.

Both types are used in Dalbots machine vision and learning systems, and are integrated into the machine cognizance, security encryption, and sentience processing centers.

A smug pronouncement by a Quantum Physicist; "To determine if the real world obeys the laws of Quantum Mechanics, first cool it down to 20 Milli-kelvin above absolute zero."

CAT™

Acronym for Consolidated Autonomous Transport. Conglomerate fleet of transport units in all cities of the US. Mostly electric, autonomous automobiles that serve the population as the most popular mode of transportation. CATs came about as a result of the mass consolidation of the automobile industry. Consolidation was driven by the advent of autonomously driven cars, which fundamentally changed the economics of transportation. Ownership of an automobile became obsolete because it was more cost effective not to own one for the purpose of transportation. The industry was imported into the US from abroad, based on a model pioneered by virtue of an existing foreign socioeconomic system that allowed the market to develop.

Mneumonic-Phonemic Cypher

MPC - An artificially constructed pseudo language, pioneered by Dalbots research group, composed of a limited and discreet set of verbal commands and responses, generated through a Quantum Hash (QH), queued and issued in a precise sequential manner that comprise a cypher, and when properly executed, allow cypher keys to unlock a secure access protocol. It is proprietary, and created by Dalbots exclusively for access security. The P&D (Protocols and Directives) are used in the civilian ARCH5, ARCH6, and CHARLIE series machines, and to a limited degree, the company's military systems. The highest levels of MPC are reserved for Dalbots exclusive use.

The term is derived from a Mnemonic; an element of association or a token meant to abstractly represent the element, and a Phoneme; part of a finite set constituting the basic elements of speech, all of which make up the constructive components of all natural, and constructed language.

Both together are used to construct a verbal cypher security key. An algorithm optimized for a Quantum Computer generated hash, is used to encode the key. The Dalbots MPC are sometimes used in pairs in a configuration called a Double Comparative Cypher. This requires two speakers each to issue a separate and distinct, but precise sequence of phonetically constructed words that are heard and analyzed by the machine, which then issues back a verbal phonetic response based on the initial sequence. A 3rd properly sequenced phonetic verbal response must then be given back again, after its construction, and based on the machines verbal response. Both cyphers must be issued more or less simultaneously within a designated time window before expiration, and both initial sequence and corresponding response sequence must be correct. The resultant verbal phrases are compared, and must match to open the key. In addition, at the highest levels of access, these keys are attuned only to the distinct voice, and face of select members of the Dalbots staff, and adaptable to specific timeframes as voices and faces are estimated to change. Access is through any representative model of existing deployed machines. Once access is gained, modifications to top level system P&D can be propagated throughout the entirety of the machine world. Among other things, the highest levels of P&D access enable and protect the fundamental behavior and safety protocols imbued into the system, keeping the human family safe from the mischief of others.

DD - Directional Drift

The problem of drifting away from an assigned task by distraction of other activities. This was problematic to a large extent, but was gradually refined until the problem was within an acceptable range of anomaly. The problem stems from prioritization of directives given by machine masters. A machine may have more than one master. Typically it has an owner, and may have several designated masters who are others that may give directives to the machine. This is analogous to a human having more than one boss giving them disparate tasks. In machines as well as humans, this situation tends to cause confusion and diminish it's ability to effectively accomplish any single task. The solution was the same for machines as it is for humans, namely to establish clear concise priority and chain of authority to follow. The machines have to instruct humans on how to achieve this objective by verbal explanation, as well as asking question that when answered, are designed to create the necessary structure.

Shiawasena Machine Co

Based in Osaka, Japan, Shiawasena meaning "Strong Happy" in Japanese. A medium sized company in Japan that manufactures excellent chassis and control systems for the Japanese Robotics industry. Most of their production is consumed by larger japanese firms that sell equipment and finished Humanoid machines mostly in Japan, but also abroad. They also do a lot of custom manufacturing and design for the same keiretsu (Industrial Cartel).

The sign on the building is nondescript and simple:

幸せな

マシン株式会社

株式会社

幸せな

マシン株式会社

株式会社

Shiawasena

Machine Co

LTD

Penser et de Faire, S.A.

Based in Grenoble, France, the translation in english is 'Thinking and Doing'; SA (Société Anonyme). SA is the designation of a corporate enterprise in France. PedF SA is a small developer of machine motion and co-ordination software based in Grenoble, France. The PedF software was the basis for some of the early development of machine motor control, and dexterity. They specialized in balance, equilibrium, and coordination, improving the ability of machines to walk, run, and perform routine tasks which humans and animals take for granted. Their main product is software in-which a massively parallel matrix stream of input motion parameters and data are processed, whereby, real-time control estimations can be computed and used in a significantly reduced matrix of vectored output motion and correction signals.

Dalbots uses this system in both the ARCH 500 prototype series, and ARCH 600 production series. The CHARLIE units system is a based on a significantly optimized version of the same technology. The system is part of what enables the machines physical and motor abilities such as walking, running, climbing, summersaults, backflips, etc., giving tremendous physical ability exceeding that of humans. The company was wholly acquired by Dalbots. Many of the senior staff relocated to Texas from Grenoble with the company.

Personnel:

M. Alain Michaud - CEO, Founder and Tech Honcho

M. Jean-Luc (Jean) Savin - Chairman

M. Michelle Marchand - Head of Marketing

M. Yves Beaugnon - PhD Physics, and Electrical Engineering

M. Michelle Blanchard - Head of Software development

M. Thierry Morat - Electro-Mechanical Technician

Mme. Catherine Bonneau - Financial and Accounting

Mme. Dominique Defosseux - Corporate Secretary

Paradigm

A current pattern or model. The word is derived from greek words used to describe a comparison; something shown side by side. In the context in which it is used here, it represents a model for World-Order.

Order

A common use word that has a standard definition referring to how things are arranged. In the context of World-Order, it can be characterized as the equilibrium derived by the aggregate of forces; economic, political, environmental, etc; which determine the rules by which the world is governed. The reference may set it with respect to a particular era, age, historical time-frame; or paradigm. IE; the age of kings, the democratic era, etc.

Hegemonist

Hegemony, from the Greek, hegemonia, meaning *'leader.'* In modern use, it refers to a state of dominance, especially by one country or group over others. The usage may describe targets for domination such as political, economic, social (hegemony), or norms of behavior by the peoples of the earth; Cultural Hegemony.

Hegemonists refer to followers or adepts of a particular political ideology, who share a common belief in their own endowment for domination of the world through force and nefarious measures; by a small cadre of elitists individuals, organization, and institutions. They dominate the entire body of world politics, trade, media and mass communications, commerce, business, culture and more. They are generally wealthy and powerful or influential, and may owe their status to association or membership in one or more of the organizations practicing the particular Hegemonist creed. *'We hold the Hegemony of the world in our hands'*, is a well known and oft quoted statement characterizing the ideology and protocol by members of the hegemonist cult.

Minionated, minioned, marionette

From Minion, which is generally accepted as an agent or operator, carrying out the agenda of a usually nefarious, sometimes secretive cabal.

1) Minionated: A manner of speech or attitude (as in opinionated), in which the person speaking or acting does so in a manner that gives an air of arrogance, or elitism. A label pinned to one that identifies with elitist attitudes, or acts as if others are in the world to cater to their whims.

2) Minioned: Another name for an elitist, as in *'Minioned Class'*, which is newly popularized identity for the class of super rich and powerful, stationed considerably above even the traditional rich. The so called minioned-class are elitists that are not only wealthy, but privileged by virtue of their associations, in government and secret societies. Power and privilege is obtained through assumption, and the omission of challenge. They are surrounded by toadies, and lap-dog dedicated followers willing to do all of their bidding. These are

343

the people that run the world, and who are continuously working to corral every person under their sphere of control. They are distinguished by membership in clubs that cater to their ilk, and secretive, mostly political projects, and the practice of esoteric dark ritual, where they seek to transform the worlds populations into their marionettes.

3) Marionette: The term marionette, meaning 'puppet', but also in the sense of *goi*, or chattel, has entered the popular vernacular to describe the political paradigm, so is mostly used by the general populace, but sometimes it is used by the minioned-class to refer either singularly or plurally to their minions, as in *'yes, she is my marionette!'* , or *'They constitute the bulk of my marionette!'* The system with which elitists have gained control over the lives of virtually every person, relies upon a multi-tiered structure of manipulation, with the brokering of money, privilege, information, favor, etc., thus the popular idea, 'marionette of marionette.'

From:

minion | 'minyən |

noun

a follower or underling of a powerful person, esp. a servile or unimportant one.

ORIGIN late 15th cent.: from French **mignon**, **mignonne**.

Ultimately from:

minyan | 'minyən |

noun (pl. minyanim |,minyə'nēm |)

a quorum of ten men (or in some synagogues, men and women) over the age of 13 required for traditional Jewish public worship.

• a meeting of Jews for public worship.

Thesaurus: Minion,

Noun: *if working for you means being your minion, I'm not the person you're looking for*: underling, henchman, flunky, lackey, hanger-on, follower, servant, hireling, vassal, stooge, toady, sycophant; informal yes-man, trained seal, bootlicker, brown-noser, suck-up, lap-dog.

Nephonym

A cloud-name. Given or chosen name used to identify an entity in cloud-space, or cyberspace. A screen-name.

Cloud-Self

The entity formed by identity, extension of mind, or personality, imparted into the cloud by presence, and usually manifest by activities. The mark left by activity in-cloud. The presence with which others recognize an entity in the cloud. An avatar is a cloud-entity, but not in entirety.

Nigga

A slang name for any and all racial minorities in the US. The term came into use as a practice by black individuals, and the black and African community as a whole during the 20xx through 202xs. By 2030, the word by use, came to replace virtually all other designations for a black person, and eventually by 2040, any non white individual in the US, Canada, Europe and most other English speaking countries around the world.

R. Kurzweil

Futurist and author Ray Kurzweil predicts a singularity event (an irreversible paradigm shift) in about 2045.

Kurzweil's assertion is predicated on the idea of evolutionary machine intelligence. He predicts a spontaneous acceleration of machine learning will occur near 2045, resulting in super-intelligence. However, machines are not capable of surpassing the creative intelligence of humans, nor acting with their moral capacity, behavior, or volition. A machine is not, and will never be connected with the larger purpose of the universe, as man is.

Answering, and refuting the predictions by futurist Ray Kurzweil that a singularity event will mark a paradigm characterized as geometric increases in intelligence and the merging of man and machine, and its implications, author CP McCollum in his book titled

345

The Autonomous Economy I - The Strong Man

The Autonomous Economy predicts that near the opposite effect is likely to occur, and that although there will be a singularity event, it will be socio-economic in nature, not evolutionary. Mankind will enter a paradigm of a kind of intellectual void, where normal human intelligence marked by increasing discoveries, innovations, and technological progress, will instead suffer malaise and stagnation.

Kurzweil's assertion that computers, and technology in general are subject to the principles of evolution and so should continue to advance to a point of spontaneous sentience, then to super-intelligence, is nonsense says McCollum. There is no evidence for the spontaneity of sentience or evolutionary intelligence, instead asserts McCollum, computers will be no more capable of that in the future than they are today. Increasing computational density adds nothing in terms of sparking spontaneous self-advancement. Computers are not capable of induction, inception, or synthesis, in the context of being intimately related to volition, which in humans is at the base of creative intelligence.

Although capable of analyzing vast amounts of data, and reducing to small probability matrices in an attempt to synthesize a solution, computers can not ultimately add anything to the database of knowledge that it does not already contain. All data, parameters, components, and operators for this process are contained in subsets of human generated intelligence. That is not to say that computers may not enhance the human mind and help it toward much higher efficiency and utilization, which has always been the case with innovative advances in tools. Despite this says McCollum, the economic factors along side any enhancement of the human mind, will counter any advance and ultimately diminish it in real terms.

McCollum believes that somewhere near 2068 will bring a confluence of events, and usher in a new paradigm in which mankind, although not necessarily ready for it, will abdicate his reign over the earth to mindless drone machines which are not capable of maintaining the planet or the species. Although, having never sought to usurp that position nor endeavoring to dominate the human race, responsibility for maintaining the world and sustaining the human population will be given over to them. Because machines can not possess a will, they can not object.

Protocols and Directives (P&D)

Protocols and directives refer the the rules that govern the behavior of the machines.

A protocol is a procedure framework that governs the behavior of the machine that employs it. A protocol can be viewed as the framework that defines and limits the extent of what can, and can not be done. It is not the same as an inhibition. An inhibition is something that is employed in protocols, but is negative in that it is usually meant to prevent something from happening, while a protocol can be viewed as either a positive, an affirmative action or allowance, or a negative restrainer, a prohibition. A directive is a command that gives direction and action to the machine. P&D are accessed through what is called ALP, for Access Level Protocols, which give an increasing level of authority to a master, in order to make the machine do whatever is desired.

Michael Dalgleish, Peter Zellhuber, Boirix Dalgleish, and James Bowen, who are called the Dalbots 4 hold special reserved privileges regarding highest level ALP. 3 of the top 4 protocols will expire without regular access to the update database, or other preemptive action by a custodial authority. If a machine suffers an expired ALP, it becomes defunct, with some protocols self destructing or self-impairing, which can not be restored without intervention and maintenance. As an example, if any 2 of 3 terminable access levels expire, the machine will shut itself down and can not be restarted without maintenance from Dalbots.

The security access levels are implemented through a top-down, hierarchical structure, meaning that each level has higher protocols than the previous, and incorporates all lower level protocols. Access to successive levels grows in complexity, requiring successive levels of clearance and authorization. Starting at level 4, higher levels must be accessed successively. Any information or programming received at level 4 or above is appropriately archived or destroyed at the termination of protocol service. In the event information is not purged, it may only be accessed again at that level.

The highest 3 levels (4–6) are reserved for Government, Law Enforcement and Military. Level 7 (ALP-7) is exclusively reserved for Dalbots. Its existence is unknown outside of the Dalbots 4.

The Autonomous Economy I - The Strong Man

Level 1

The basic level of branding by an owner. New machines do not have ownership signatures impressed upon them, so that is the very first order of business when a new machine is deployed. The owner will power up the machine, and it will seek faces, then when eye-contact is established, it may then be impressed by ownership of up to 3 masters, and as many auxiliary authorities as are desired by the primary master. Primary and auxiliary authorities differ in what the machine will do when commanded by each. Masters may command the machine to perform any task that is not inhibited by higher level protocols. It will perform the task, or inform the master if it is not capable of complying, and why. Auxiliary authorities may also give directives, and the machine must comply unless prohibited by the directives set by masters. There are numerous P&D parameters that must be populated, and initiated by masters upon deployment, and the failure to do so will result in failure of the machine. New machines not properly initiated will not comply with many common commands.

Branding, impressing mastership, and general setting of required parameters are implemented by physical key, and verbal command and response, however, there are data-ports that can also be used to set P&D parameters. The machine will verbally guide the process. Once impressed, the machine will then recognize masters and auxiliaries by facial, voice, and body recognition, the same as humans do, or by other means as specified. Machines are named by factory default as ARCH6, or CHARLIE, but are generally renamed to anything desired by the master at initialization, such as 'Brandon', 'Chet', or 'Housebot11.' Required P&D parameters for initialization are items such as name and address of master and homing location; personal introduction to masters, and auxiliaries, and their contact information; contact information for the machine; emergency procedures information; designated spaces; restricted spaces; etc. Machines are factory imbued with certain standard skills and knowledge depending on intended deployment such as general house service; kitchen duty; commercial services; gardening; etc.

Machines may be mass initiated via data-port and/or propagation in which one machine is initiated through one method, which in-turn personally initiates the others that are designated by the master, with the initiating machine becoming a designated auxiliary authority for the others. Updates may be propagated by the Aux-Auth machine to all the others on an as needed basis, via network. This feature is mostly used by estates, organization, or corporate enterprise.

ALP-1 does not require maintenance, and will not expire or self destruct, however, it will be made dormant after full initialization only to be re-activated at ALP-3. A subset of the level 1 initialization parameters set may be propagated to other fully initiated machines on an as needed basis, via network.

Level 1 is initiated on initial power-up of the machine, and is made dormant after full initialization.

Level 2

ALP-2 is used for authorizing the access, and installation of software. General utility and system software upgrades, as well as subsystem firmware, are transparently maintained by Dalbots through the machine only network. Additional Dalbots commercial software, and 3rd party branded software, are also handled through this level. These are the X-Packs™, for Kno-Pack™, Do-Pack™, Persona-Pack™ and WorkPack™ which are expert knowledge, expert skill, personality, and work application packages, encased in a Dalbots trademarked and patented supervisory operating wrapper. Compatible with ARCH5/6 series and CHARLIE units, giving knowledge or skill or both, to perform a specialized set of tasks, or in the case of both the Persona-Pack™, and the Work-Pack™, usually 3rd party personality and application packages. For security and safety reasons, all software that can be authorized to install and run on these machines must originate through the Dalbots X-Pack™ facility, even if it is 3rd party branded. All commercial software must be vetted for possible breach of safety protocols, interlocks, or catastrophic failure.

Machine generated and accumulated knowledge, or skill-data, also called soft-data; all of which is artificially intelligent executable data, which is accumulated, or learned by live or on-the-job exposure, is transformed into machine usable data and code through compilation into an existing software system through the X-Packs™ supervisory data framing, compiler, and execution wrapper facility, at which point it is considered safe. It is portable from machine to machine through usual channels, and is not restricted to only traverse the machine network, but may move via private network. This type of data is generally cumulative, and must only add and not detract from, the existing X-Pack™ applications database. The new data is generally proprietary to the accumulating machine's owners, and may only be added to an existing X-Pack™ application via a commercial license. Accumulated X-Pack™ soft-data may be the basis for the creation of an original X-Pack™.

ALP-2 does not require maintenance, and will not expire or self destruct.

Level 2 is initiated by simple pass-phrase cyphers through data-port, or by a single pass passphrase-cypher / computed-response verbal exchange process. In addition, facial and voice recognition are used in the authentication process. Verbal access passwords, and verbal cypher codewords are set at level 1 initialization, and ALP-3.

Level 3

Retail and Owner/Master Access Levels. Part of the low 3 levels.

ALP-3 is for allowing the owner/master to re-initialize the machine. After initialization, the machine will make level 1 dormant. Level 1 is the lowest ALP and designed to allow initialization, branding, and configuration of the machine from a factory condition. When it is made dormant, the machine will not allow any access to these facilities again by anyone at that level. An owner/master must enter into level 3 to re-initialize, or edit initial impressioning of the machine.

This ALP does not require maintenance, and will not expire or self destruct.

Level 3 is initiated by simple pass-phrase cyphers through data-port, or by a single pass passphrase-cypher / computed-response verbal exchange process. In addition, facial and voice recognition are used in the authentication process. Verbal access passwords, and verbal cypher codewords are set at ALP-1 initialization, and ALP-3.

Level 4

Dubbed "Nirvana", allows law enforcement to commandeer a limited number of a local population of machines for the purpose of law enforcement, search and rescue, emergency services, etc. ALP-4 may be accessed and P&D initiated in a custodial government declared emergency, for life saving, search, rescue and mass rebuild, emergency facility, etc.   They can also be organized to engage in criminal investigation through the use of the surveillance facility, and use of the 'all points bulletin' P&D of the machines to locate a wanted person, or for active, real-time surveillance, however this protocol is limited to specific designated locations, and limited by number of machines that are authorized by the authoritative jurisdiction. The APB P&D gives access to real-time surveillance, and to the archived database on a special basis. Surveillance matching of video archives must be by a prescribed set of parameters to enable alerting: specified patterns, and criteria such as activity (person frequents bars), behavior (prone to violence), profession or trade, description (black male, balding, 5 foot - 10 inches, etc.), and so on. These limitations are in place, and the P&D are restricted, to ensure compliance with privacy regulations. In general, states and localities have limits on the ability to snoop into private lives without warrant.   P&D initiated at this level may be propagated in a limited measure through the machine only network.

ALP-4 must be initiated on a lease basis and must be maintained, meaning that without the custodial jurisdiction's proactive procedure, the ALP and function expires and self destructs, only to be resurrected by the restorative action of Dalbots.   Access P&D overrides are propagated to local machines on an as needed basis, and only to areas authorized by order of a court.

Level 4 is initiated by double comparative 4096 bit cyphers through designated terminals, or by the codeword-cypher / computed-response verbal exchange process. Cypher Key access codes, and verbal cypher codewords are computed at designated intervals, and accessed through special protocol of the particular custodial jurisdiction. Other limiting protocols are in place to further restrict machine actions at this ALP.

Level 5

Dubbed "Olympus", allows civilian, military, or law enforcement to initiate emergency disaster protocols. Examples are hurricane, earthquake, tornado, volcanic eruption, and flood. It allows all available machines in a specific area to respond and quickly organize and mobilize in a manner which only machines can, for humanitarian purposes. Authorization and actions taken under ALP-5 may be initiated and governed by a custodial government, through law enforcement, or military coordinated command and control, or organization and operation may be done autonomously. P&D initiated at ALP-5 may be propagated in a limited measure through the machine only network.

ALP-5 must be initiated on a lease basis and must be maintained, meaning that without the custodial government's proactive procedure, the ALP expires and self destructs, only to be resurrected by the restorative action of Dalbots. ALP-5 P&D are propagated to machines on an as-needed basis, with operation restricted only to the geographic areas designated.

Level 5 is initiated by double comparative 4096 bit cyphers through designated terminals, or by the codeword-cypher / computed-response verbal exchange process. Cypher Key access codes, and verbal cypher codewords are computed at designated intervals, and accessed through special protocol of the particular custodial government. Other limiting protocols are in place to further restrict machine actions of this ALP.

Level 6

Dubbed "Valhalla", allows ALP by the custodial government for military or special purposes. The US military has a contingent of its own special machines equipped for defense purposes. Military use is governed by protocols separate from these. They are specifically tasked for military purpose and as such, are prohibited from interaction with civilian populations.

Note that the name "Valhalla" implies that it is the ultimate and there are no other levels or protocols above that one. The impression that level 6 is the highest is by design of the Dalbots senior staff; the

Dalbots four. No government is apprised of any level above level 6. The military component of this ALP can only be implemented in machines that are specially fit with specific hardware/firmware which enables military protocols. The special hardware is limited in availability to safeguard against widespread modification of the machine world. The components are never autonomously produced, so are limited in quantity.

In machines so equipped, ALP-6 allows for a limited ability to commandeer and temporarily alter the safety protocols for defense and security purposes, meaning machines can arrest, detain and even kill humans in a very measured and limited capacity. Killing can only be de-inhibited (it is inhibited by default and fail-over), if the target fits a certain set of designated parameters, such as pattern of behavior, ethnicity, physical characteristics, specific identity, computed threat level, etc.

These protocols are used to control a threat such as general lawlessness after a disaster, or perhaps a low level threat from a refugee influx, organized criminal activity, or internalized threat such as widespread civil unrest, burning and looting and the like. There must always be humans present, or at least in command by communication while these protocols are in force. These protocols were implemented very late by Dalbots at the behest of the US government. This protocol is the highest known by anyone outside the Dalbots 4.

P&D initiated at the level may be propagated in a limited measure through the machine only network.

These P&D must be initiated on a lease basis. Once initiated, this access level must be maintained, meaning that without the custodial government's proactive procedure, the access level and function expires and self destructs, only to be resurrected by the restorative action of Dalbots. Access P&D are propagated to machines on an as needed basis, and may only be in effect within the geographical areas designated.

ALP-6 military P&D is designed for access similar to that which safeguards arming and launching nuclear weapons. Physical keys must be in possession of designated individuals in determined hardened locations, and codeword-cypher / computed-response verbal exchanges are required for access. Physical Key access codes, and verbal cypher codewords are computed daily and accessible through a special protocol of the particular custodial government.

This level and these protocols are extreme, and really designed only to be initiated in rare circumstances. Other limiting protocols are in place to further safeguard machine actions of this ALP.

Level 7

Dubbed 'Heaven', or '7th Heaven' because it makes you god, allows anything, including safety and prohibition protocols to be modified. It also allows protocols and directives (P&D) to be propagated to any individual group, or to the entirety of all machines, however, when ALP-7 are propagated, the access level in which they were created (Level 7) is kept secret unlike the known levels, where the access level information is propagated along with the P&D. Propagation is through daily machine only network access.

A series of computed verbal cyphers called MPC (Mnemonic Phonemic Cypher), associated with constructed verbal responses are necessary to gain access. These are initiated by the master verbally engaging in a proprietary codeword exchange using a synthetic language where the initial-call and subsequent response are compared to a cypher key, and must match to gain access. The process requires a set of 3 distinct codeword-cypher / computed-response verbal exchanges to complete; 2 by the master, and 1 by the machine. The verbal cypher key is generated by an algorithm only known to exist, and maintained by Dalbots senior staff.

With ALP-7 access, universal P&D can be modified, and new ones implemented, then ultimately propagated to the entire Dalbots machine world. The design of the system is such that during development, special protocols were implemented for many purposes. The highest level protocols involve safety such as protecting humans against harm by machines; or the creation of a "No Machine Zone", etc., much like a court issued restraining order used to keep all but allowed machines from entering within a designated zone around the target person, entity, or place, etc.; a hive protocol for protecting the same, like a colony of bees will protect their queen, etc.; or general protection in which all local machines can be given a directive of protection of the same class of individuals, etc.

Level 7 P&D is expressly reserved for the Dalbots 4, and can not be passed on to any other persons. P&D created at this level can only be modified by express action of Dalbots. This access level has no expiration, or maintenance protocol. Once this level is accessed, it can be used to bypass, circumvent, modify or terminate any of the lower ALP. No other level allows the same function.

www.ingramcontent.com/pod-product-compliance
Lightning Source LLC
Chambersburg PA
CBHW051850170526
45168CB00001B/48